Industrial Automated Systems: Instrumentation and Motion Control

Industrial Automated Systems: Instrumentation and Motion Control

Contributors

Lei Wang, Muguo Li et al.

AURIS
Reference

www.aurisreference.com

Industrial Automated Systems: Instrumentation and Motion Control

Contributors: Lei Wang, Muguo Li et al.

Published by Auris Reference Limited

www.aurisreference.com

United Kingdom

Industrial Automated Systems: Instrumentation and Motion Control

ISBN: 978-1-78154-937-7

British Library Cataloguing in Publication Data
A CIP record for this book is available from the British Library

Printed in the United Kingdom

Exclusively distributed by CBS Publishers & Distributors Pvt. Ltd.

Sales & Distribution Rights only for India, Pakistan, Bangladesh, Sri Lanka, Nepal and Bhutan. This book is not to be sold outside these territories.

Contents

List of Abbreviations

A/D	Analog-to-digital
aPS	Automated production systems
ADSs	Automation Sequence Diagrams
CG	Centre of gravity
CNF	Composite nonlinear feedback
CT	Computed tomography
CNC	Computer numerical control
CE	Conformité Européenne
DHP	Data Highway Plus
DOF	Degree-of-freedom
DA	Delmia Automation
DCS	Destination Control System
DC	Distributed clock
DDE	Dynamic Data Exchange
EMF	Eclipse Modeling Framework
ED	Entity Diagrams
ESC	EtherCAT slave controller
ETG	EtherCAT Technology Group
FDA	Food and Drug Administration
FCS	Frame Check Sequence
FBD	Function Block Diagram
GB	Gigabyte
GPS	Global positioning systems
HiL	Hardware-in-the-loop
HMI	Human machine interface .
IMU	Inertial measurement unit
IO	Input/output
IL	Instruction List
KB	kilobytes
LD	Ladder Diagram
LMI	Linear matric inequality
LPV	Linear parameter-varying
MII	Medium Independent Interface
MDD	Model-driven development
ML	Modelica
MiL	Model-in-the-loop
OGCC	optimal guaranteed cost control
OEE	Overall Equipment Effectiveness
PPU	Pick and Place Unit
PiL	Processor-in-the-loop
PLC	Programmable Logic Controller

QFT	Quantitative feedback theory
RAM	Random access memory
RTC	Real-time clock
RTK	Real-time kinematics
RAMI	Reference architecture
ROS	Robot Operating System
SPI	Serial Peripheral Interface
SMC	sliding mode control
SMC	Sliding Mode Control .
SiL	Software-in-the-loop

List of Contributors

Lei Wang
School of Computer Science and Technology, Henan Polytechnic University, Jiaozuo 454000, China
State Key Laboratory of Coastal and Offshore Engineering, Dalian University of Technology, Dalian 116024, China

Muguo Li
State Key Laboratory of Coastal and Offshore Engineering, Dalian University of Technology, Dalian 116024, China

Junyan Qi
School of Computer Science and Technology, Henan Polytechnic University, Jiaozuo 454000, China

Qun Zhang
State Key Laboratory of Coastal and Offshore Engineering, Dalian University of Technology, Dalian 116024, China

Daniel K. Fisher
USDA Agricultural Research Service, Stoneville, USA

Peter J. Gould
US Forest Service, Pacific Northwest Research Station, Olympia, USA

Birgit Vogel-Heuser
Institute of Automation and Information Systems, Technische Universität München, Munich, Germany

Susanne Rösch
Institute of Automation and Information Systems, Technische Universität München, Munich, Germany

Juliane Fischer
Institute of Automation and Information Systems, Technische Universität München, Munich, Germany

Thomas Simon
Institute of Automation and Information Systems, Technische Universität München, Munich, Germany

Sebastian Ulewicz
Institute of Automation and Information Systems, Technische Universität München, Munich, Germany

Jens Folmer
Institute of Automation and Information Systems, Technische Universität München, Munich, Germany

Nguyen Hoang Giap
Department of Intelligent System Engineering, Graduate School of Dong-eui University, Busan, Korea

Jin-Ho Shin
Department of Mechatronics Engineering, Dong-eui University, Busan, Korea

Won-Ho Kim
Department of Mechatronics Engineering, Dong-eui University, Busan, Korea

Ngo Ha Quang Thinh
Department of Intelligent System Engineering, Graduate School of Dong-eui University, Busan, Korea

Jae-Gark Choi
Department of Computer Engineering, Dong-eui University, Busan, Korea

Won-Ho Kim
Department of Mechatronics Engineering, Dong-eui University, Busan, Korea

Timo Vepsäläinen
Department of Automation Science and Engineering, Tampere University of Technology, Korkeakoulunkatu 3, 33101 Tampere, Finland

Seppo Kuikka
Department of Automation Science and Engineering, Tampere University of Technology, , Korkeakoulunkatu 3, 33101 Tampere, Finland

M. K. Aripin
Control, Instrumentation & Automation Department, Faculty of Electrical Engineering, Universiti Teknikal Malaysia Melaka, 76100 Durian Tunggal, Melaka, Malaysia

Yahaya Md Sam
Department of Control & Mechatronics, Faculty of Electrical Engineering, Universiti Teknologi Malaysia, 81310 UTM Johor Bahru, Johor, Malaysia

Kumeresan A. Danapalasingam
Department of Control & Mechatronics, Faculty of Electrical Engineering, Universiti Teknologi Malaysia, 81310 UTM Johor Bahru, Johor, Malaysia

Kemao Peng
Temasek Laboratories, National University of Singapore 5A Engineering Drive 1, Singapore 117411

N. Hamzah
Faculty of Electrical Engineering, UiTM Pulau Pinang, 13500 Permatang Pauh, Pulau Pinang, Malaysia

M. F. Ismail
Industrial Automation Section, Universiti Kuala Lumpur Malaysia France Institute, Section 14, Jalan Teras Jernang, 43650 Bandar Baru Bangi, Selangor, Malaysia

Erwin Normanyo
Department of Electrical and Electronic Engineering, Faculty of Engineering, University of Mines and Technology, Tarkwa, Ghana

Francis Husinu
Accra Brewery Limited, Accra, Ghana

Ofosu Robert Agyare
Department of Electrical and Electronic Engineering, Faculty of Engineering, University of Mines and Technology, Tarkwa, Ghana

Sagarika Pal
Department of Electrical Engineering, National Institute of Technical Teachers Training and Research, Kolkata. [Under MHRD, Govt. of India], Block-FC, Sector-III, Salt Lake City, Kolkata-700106, India

Niladri S. Tripathy
Department of Electrical Engineering, National Institute of Technical Teachers Training and Research, Kolkata. [Under MHRD, Govt. of India], Block-FC, Sector-III, Salt Lake City, Kolkata-700106, India

Luis Emmi
Centre for Automation and Robotics (UPM-CSIC), Arganda del Rey, 28500 Madrid, Spain

Mariano Gonzalez-de-Soto
Centre for Automation and Robotics (UPM-CSIC), Arganda del Rey, 28500 Madrid, Spain

Gonzalo Pajares
Department of Software Engineering and Artificial Intelligence, Faculty of Informatics, University Complutense of Madrid, 28040 Madrid, Spain

Pablo Gonzalez-de-Santos
Centre for Automation and Robotics (UPM-CSIC), Arganda del Rey, 28500 Madrid, Spain

Yuqiang Luo
Shanghai Key Lab of Modern Optical System, Department of Control Science and Engineering, University of Shanghai for Science and Technology, Shanghai 200093, China

Zidong Wang
Department of Computer Science, Brunel University London, Uxbridge, Middlesex UB8 3PH, UK
Communication Systems and Networks (CSN) Research Group, Faculty of Engineering, King Abdulaziz University, Jeddah 21589, Saudi Arabia

Guoliang Wei
Shanghai Key Lab of Modern Optical System, Department of Control Science and Engineering, University of Shanghai for Science and Technology, Shanghai 200093, China

Bo Shen
School of Information Science and Technology, Donghua University, Shanghai 200051, China

Xiao He
Department of Automation, Tsinghua University, Beijing 100084, China

Hongli Dong
College of Electrical and Information Engineering, Northeast Petroleum University, Daqing 163318, China
Research Institute of Intelligent Control and Systems, Harbin Institute of Technology, Harbin 150001, China

Jun Hu
Research Institute of Intelligent Control and Systems, Harbin Institute of
Technology, Harbin 150001, China
Department of Applied Mathematics, Harbin University of Science and Tech-
nology, Harbin 150080, China

Ryan A. Beasley
Department of Engineering Technology and Industrial Distribution, Texas
A&M University,TAMU, College Station, USA

Preface

The book Industrial Automated Systems: Instrumentation and Motion Control, is the ideal book to provide readers with state-of-the art coverage of the full spectrum of industrial maintenance and control, from servomechanisms to instrumentation. This book focuses on operation, rather than mathematical design concepts. It is formatted into sections so that it can be used for a variety of courses, such as electrical motors, sensors, variable speed drives, programmable logic controllers, servomechanisms, and various instrumentation and process classes. First chapter presents a new design approach based on EtherCAT protocol aiming at a special networked motion control system. The objective of second chapter is to introduce researchers and practitioners to potential applications of the opensource Arduino platform for implementation in research and monitoring applications. Third chapter reviews the vital elements for control system design of an active yaw stability control system; the vehicle dynamic models, control objectives, active chassis control, and control strategies with the focus on identifying suitable criteria for improved transient performances. Fourth chapter introduces a reference-pulse interpolator motion control system, which can be applied for computer numerical control (CNC) machine tools. The state of the art of fault handling in industrial automated production systems (aPS) is discussed in fifth chapter as a result of a case study analysis in eight companies developing aPS. In sixth chapter, a model-driven development process that includes support for design-time simulations is complemented with support for simulating sequential control functions. The approach is implemented with open source tools and demonstrated by creating and simulating a control system model in closed-loop with a large and complex model of a paper industry process. Seventh chapter reviews the vital elements for control system design of an active yaw stability control system; the vehicle dynamic models, control objectives, active chassis control, and control strategies with the focus on identifying suitable criteria for improved transient performances. In eighth chapter, consideration was given to the design of an HMI for an automated boiler plant which can be operated both manually and automatically by the press of start buttons. Ninth chapter provides a review of medical robot history and surveys the capabilities of current medical robot systems, primarily focusing on commercially available systems while covering a few prominent research projects. Last chapter strives to develop a system architecture for both individual robots and robots working in fleets to improve reliability, decrease complexity and costs, and permit the integration of software from different developers.

Chapter 1

DESIGN APPROACH BASED ON ETHERCAT PROTOCOL FOR A NETWORKED MOTION CONTROL SYSTEM

Lei Wang,[1,2] Muguo Li,[2] Junyan Qi,[1] and Qun Zhang[2]

[1]School of Computer Science and Technology, Henan Polytechnic University, Jiaozuo 454000, China

[2]State Key Laboratory of Coastal and Offshore Engineering, Dalian University of Technology, Dalian 116024, China

ABSTRACT

This paper presents a new design approach based on EtherCAT protocol aiming at a special networked motion control system. To evaluate this system, the testing platform of the networked motion control system for ocean wave maker is designed. The paper gives the detailed design for the testing platform including software and hardware. With message access delay and jitter, a set of experiments are done to evaluate the crucial performance for the implemented networked motion control system. By analyzing the experimental results, the systemic performance is verified. And the design approach can be widely extended to other automation application scenarios.

INTRODUCTION

Motion control, as a subfield of automation, has been finding applications in a broad range of areas such as packaging, printing, textile, semiconductor production, and assembly industries. As the number of the devices of motion control system increases, distributed networked motion control systems are desired in various manufacturing fields. This kind of system must be able to access a mass of device information from any locations in factory floor. In order to solve this problem, various serial communication networks have been designed and implemented to provide reliable and efficient communication paths for data exchange among the system components [1]. In the last two decades, hundreds of proprietary digital network communication protocols named fieldbus have been developed. The construction of networked motion control systems by fieldbus has become a hot topic. In particular, the fieldbuses have been widely used in a great deal of motion control applications for robotics, passenger cars, and aircrafts [1–3]. Nonetheless, because of the relatively low bit rates and (sometimes) changing cycle times, the fieldbuses are difficult to cope with the existing systemic needs for the transmission of a higher quantity of information with the tight timing constraints. Moreover, the software and hardware of multivendor products are incompatible [4]. It is quite difficult to connect the equipment of different manufacturers.

Due to ubiquity, high speed, simplicity, and low cost, Ethernet seems to be the most promising candidate for "the one" network technology to replace fieldbus [5]. Whereas Ethernet is not originally designed for real-time control, the most significant problem is that it is nondeterministic due to its bus arbitration scheme [6]. There are mainly three methods to modify standard Ethernet, including adding an industrial-automation specific application layer on top of TCP/IP, using the priority scheme at the Ethernet MAC layer, and intervening into the scheduling procedure of the MAC layer [7, 8]. The modified Ethernet protocols are called Industry Ethernet protocols which integrated the advantages of both the real-time capability and the merits of Ethernet. So many industrial companies and institutes have shown interests in developing Industry Ethernet protocols. And Industry Ethernet has become the de facto standard for industrial automation networks and widely used in automation and process control domain. In recent years, various Industry Ethernet protocols, such as SERCOS, POWERLINK, ProfiNet/ProfiNet IRT, and EtherCAT, have been proposed to develop networked control systems which have the control loop closed through a communication network [9–11]. Several researchers also applied the Industry Ethernet technology in motion control [12–14]. Also some analysis and evaluation on the performance of Industry Ethernet protocols have been done [7, 15–18]. At the same time, a considerable number of Industry Ethernet protocols have been introduced into motion control systems to implement networked motion control [19]. However, few researchers investigate the hybrid networked motion control systems with multiaxes servo motors and a great deal of sensors [20]. In view of the special application requirement for high real-time performance and the large amounts of information transmission, we design ocean wave maker networked motion control systems.

The system should accurately simulate ocean waves in the experimental basin and quickly get the experimental data to provide reliable basis for engineering design and scientific research from field sensors. Therefore, the system is constructed with a large number of distributed data acquisition modules and multiservo motors in the field of ocean engineering application. This system should be provided with the capability for transmitting motion control commands and feedback information with lower delay and jitter. Meanwhile, it can help acquire the high efficiency and high real-time large-capacity data more effectively. As a very popular Industry Ethernet protocol, EtherCAT protocol is selected to design the system

The rest of the paper is organized as follows: in Section 2, we briefly introduce the unique features of EtherCAT protocol; and then in Section 3, we describe the overall system architecture and the design of software and hardware. Experimental results are shown in Section 4 for testing and evaluating the system performance. The paper is concluded in Section 5.

MAJOR ADVANTAGES OF ETHERCAT PROTOCOL

EtherCAT is a popular real-time Industry Ethernet network defined Beckhoff and supported by the EtherCAT Technology Group (ETG) and currently specified by the IEC 61158, SEMI, ISO [21]. EtherCAT protocol users have explosively grown during the past seven years. As of May 31, 2010, the ETG has 1357 members from 50 countries [21]. It possesses some unique benefits that make it cope with networked motion control needs. Generally speaking, the major advantages of EtherCAT protocol include high data transmission efficiency and speed and high accuracy clock synchronization. Section 2 briefly presents the main merits and functions of the protocol.

High Transmission Efficiency and Speed

EtherCAT Network Structure

The whole EtherCAT network is made of master station nodes and slave station nodes as shown in Figure 1.

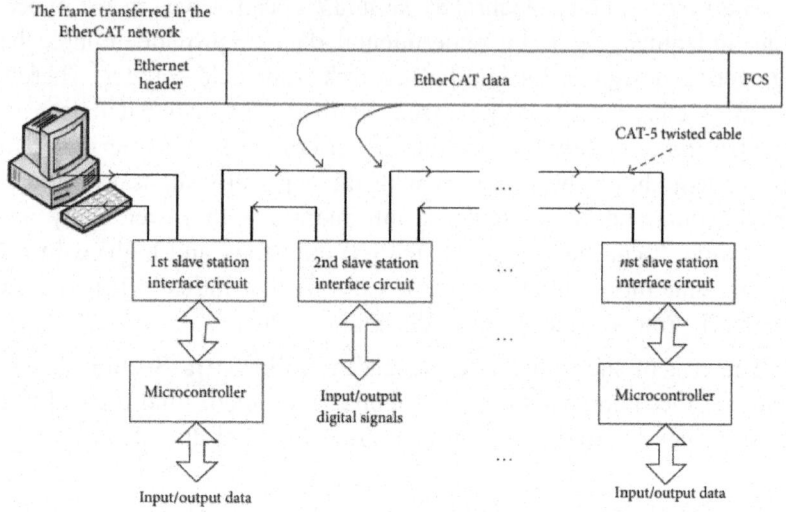

Figure 1: Whole structure of EtherCAT network with two slave stations.

One EtherCAT master (e.g., industrial PC or embedded microcontroller) is connected with a certain number of EtherCAT slaves [13]. All stations are connected together to create a logical ring through standard Ethernet cable. The master station sends the standard Ethernet frames with EtherCAT data around the ring. When the frames pass through each slave station node, they are processed without stopping. At the open end, the frames are transferred backward to the master station.

Structure of EtherCAT Frame

In EtherCAT protocol the basic Ethernet frame structure is not changed. Thus, EtherCAT is compatible to other Ethernet protocols. EtherCAT frame can be distinguished from other Ethernet frames through the Ethernet type field with the special type identifiers 88A4. A standard Ethernet frame consists of five fields: preamble, start of frame delimiter, header, data, and a Frame Check Sequence (FCS), as in Figure 2.

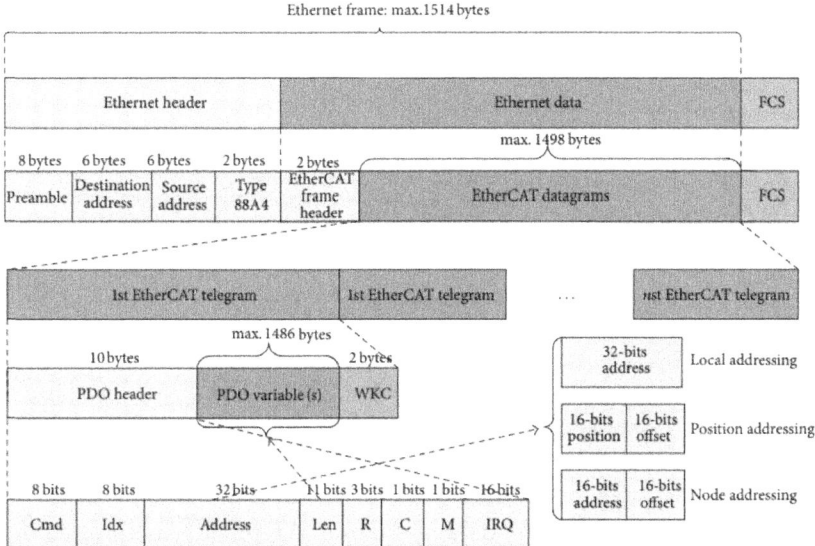

Figure 2: EtherCAT frame structure.

Specific Hardware of Slave Station and Addressing Image

Slave station node is mainly made up of an ASIC chip known as EtherCAT slave controller (ESC) and an application controller. The ESC implements the communication interface function between EtherCAT network and slave application. Each ESC possesses a consistent random access memory (RAM) which is used to exchange data between the EtherCAT master and local slave application controller. The special data link layer hardware units, named Synchronize Managers (SMs) and Fieldbus Memory Management Units (FMMU), carry out the management of the user memory of ESC and the addressing images from EtherCAT PDOs to physical memory units. Figure 3 shows how FMMUs are configured to map physical memory to PDOs images.

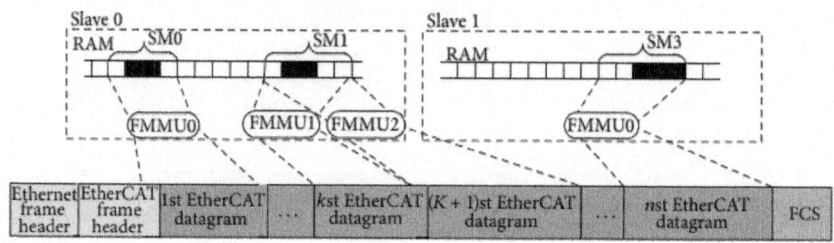

Figure 3: Management of RAM using SMs and FMMUs.

Each PDOs image embedded in EtherCAT frame has a unique local address which is assigned by master station. During start-up phase, the master station writes the registers of FMMUs with the values: a local bit start address, a physical memory start address, a bit length, and a type field that specifies the direction of mapping (Input/Output). The application controller of slave station extracts or inserts the data into EtherCAT frame according to the read values stored in the RAM units of ESC from the registers of FMMUs. Both the master station and the slave one need access the RAM; thus ESC must be able to inform both sides of the time when they can access the RAM; otherwise the data consistency and security cannot be guaranteed. To solve the question, the RAM units are managed by a series of SMs. Each SM controls a buffer in RAM and generates interrupt to inform the master and slave station application controller when the access of the other side has finished. SMs consist of a series of registers and they are configured by the master station during start-up phase [22].

Implementation of High Transmission Efficiency and Speed

As shown in Figure 1, when the frames from the master station pass through the slave nodes, they are not encoded or decoded. Only the data are read and written as the bits passing the ESC. After the data are accessed, the frames are automatically forwarded to the next port by the EtherCAT Processing Unit in ESC; accordingly forwarding delay only depends on the receive RAM buffer size and EtherCAT Processing Unit delay. In view of unparalleled processing technology, the EtherCAT network provides faster communication cycle time than most of Industry Ethernet networks.

Besides, Figure 2 shows that the largest EtherCAT frame size can be up to 1498 bytes; a large amount of Output/Input data to/from multislave station devices are transferred into an EtherCAT frame during one communication period. Only the relevant commands (Cmds: In EtherCAT telegram field) are recognized and executed by either any slave node or several nodes simultaneously. These commands ascertain the master accessing services for the addressable memory section of slave station nodes. Moreover, the frame may comprise the data of many devices both in sending and receiving direction, so the usable data rate increases to over 90%. For example, we can take into account a networked control system with 20 slave stations. When we visit all the stations, for other Industry Ethernet protocols, the master station must send more than 20 different packets. However, for EtherCAT, one long packet that touches all slaves is sent, and the packet contains 20 devices values of data. Besides, if all the slaves need to receive the same data, one short packet is sent, and the slaves all look at the same part of the packet as it is streaming through, optimizing the data transfer speed and bandwidth. So EtherCAT protocol possesses more high transmission efficiency and speed over other Industry Ethernet protocols. It is particularly well adapted to use in the system with a large amount of sensors.

High Accuracy Clock Synchronization

As we know, the precondition of the high performance of networked motion control is that communication protocol ensures real-time and synchronization of data transmission.

In contrast to many of other Industry Ethernet protocols, EtherCAT adopts accurate clock synchronization scheme referred to as the distributed clock (DC) functionality [23]. This function is realized by using specific hardware module of ESC and provides an identical clock time for both the slave station and master station. To present the principle of distributed clock, a line topology structure with five slave stations is depicted in Figure 4. Three different types of clocks, namely, as master clock, reference clock, and slave clock, are defined in the system [24]. The reference clock is used to synchronize all other slave stations [25]. During the system start-up phase, the master station must set the local time of the reference clock and the other slave clocks to the current reference time. To this end the EtherCAT master sends a special EtherCAT datagram at short intervals (with sufficient frequency that ensures that the slave clocks remain synchronized within the specified limits), in which the EtherCAT slave with the reference clock enters its current time [26]. This information is then read from the same datagram by all other EtherCAT slaves featuring a slave clock. The defined time parameters in Figure 4 are listed in Table 1.

Table 1: Parameters for propagation delay calculation

Parameter	Description
T_{Cx}, T_{By}	Receive time of the first bit of the measurement frame gets to the slave port (x = 1–5, y = 1–4). It will be DC Receive Time 0 register.
T_{Px}	Processing delay of slave x (x = 1–5) when the frame passes through the slave x.
T_{Fx}	Forwarding delay of slave x (x = 1–5) when the frame comes back from the slave x.
T_{Wxy}	Wire propagation delay between slave x and y (x/y = 1–5).

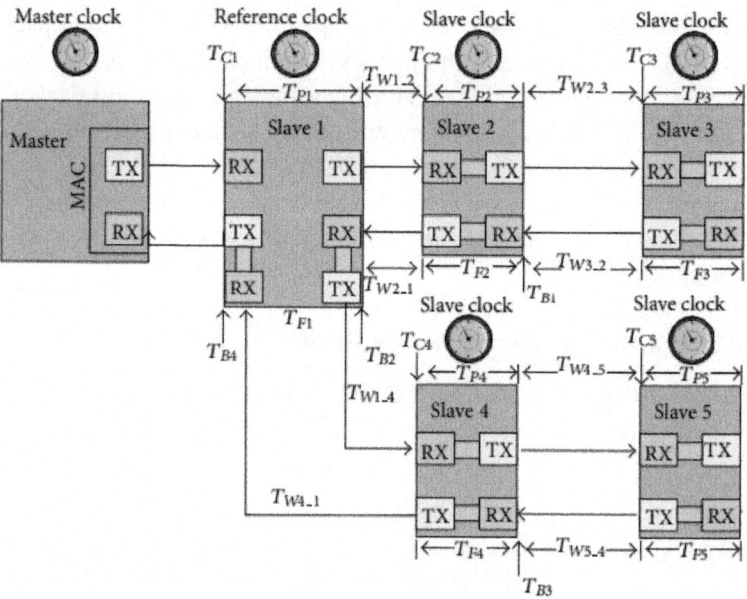

Figure 4: Principle of distributed clock.

To guarantee accurate clock synchronization, the following operation must be finished beforehand. Firstly, master station passes the propagation delay measurement frame to all slaves and the slave controller records the receive time of the frame. And the master station collects the time stamps afterwards and calculates the propagation between all slaves. Secondly, the local offset time to reference clock of every slave station must be compensated [27]. In order to do that, the time of each slave clock is compared with that of the reference clock, and the difference is stored in each slave to make the slave get the same absolute system time. Finally, the drift between reference clock and slave clock cannot be avoided because each slave station has respective free running oscillator with different parameters. With the parameters information, the master can calculate the delay of each node. The master repeats the calculation for every frame it sends. As the network operates, the enormous sample size means that the master has incredibly accurate data. The inherent ring topology creates an incredibly efficient clock mechanism that increased in accuracy with every message [13, 14].

DESIGN OF NETWORKED MOTION CONTROL SYSTEM

Principle of Networked Wave Maker Control System

We design the full closed-loop networked motion control system, as a typical application, for ocean wave maker. Ocean wave maker is a kind of experiment equipment simulating the wave generation according to calculated data under laboratory condition. The basic principle of ocean wave maker is to vibrate the water through the mechanical motion of the wave board to generate waves. The mechanical motion of the wave board should be suitable for the reciprocating smooth variable motion of the time domain wave signals. The servo should respond to changing waves rapidly with high speed and tracking precision. Due to the protocol performance limit, half closed-loop control mode is widely used in existing networked wave maker systems. In the control mode, the data of a full target wave curve are downloaded to slave station memory section, and then the servo motor executes control instructions from servo driver to control push board movement. The actual position values from encoder are feedbacked to slave station application controller. After one wave making period, the data are uploaded to master station and then the master station program modifies the next data values. This kind of control style cannot achieve complicated advanced control algorithm because of the capability limit of slave microcontroller. So the system control accuracy cannot be ensured. To improve the control accuracy of system, it is necessary to construct a full closed-loop between master station and field sensors with less than 1 milliseconds network delay. Figure 5 illustrates that the master station sends target position commands to the microcontroller via EtherCAT network. Different from half closed-loop control system, full closed-loop control mode can supply real-time field sensors data to master station before the next commands are sent. The commands are revised by the application program which runs in the master station. Therefore we proposed a full closed loop control structure between master station and field sensors. The commands are able to be revised in one communication period, so actual control curve will be more accurate.

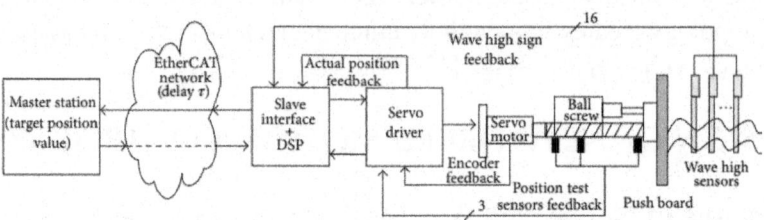

Figure 5: Principle of networked motion control system of wave maker.

Detailed Design for Networked Motion Control System of Wave Maker

In this study, we construct the networked motion control system of wave maker including 10 slave stations. Figure 6 shows the whole structure of the system.

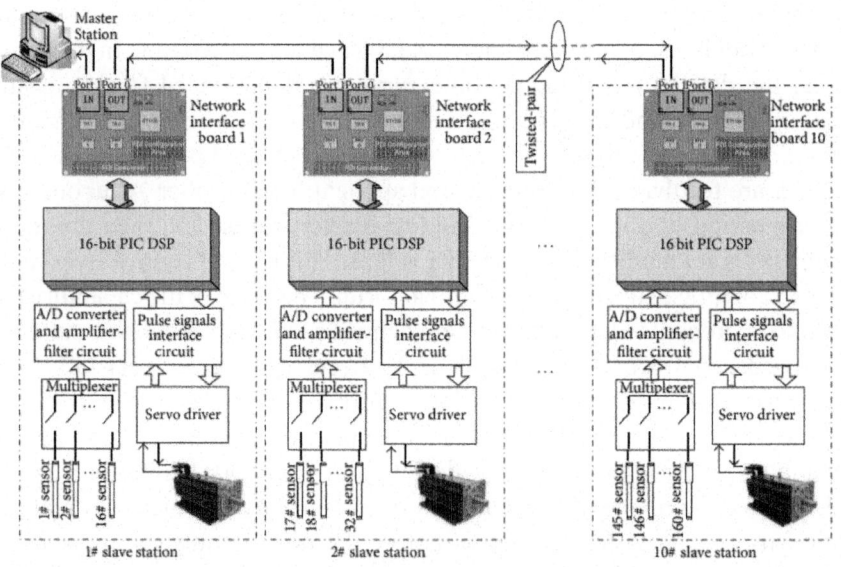

Figure 6: Structure of networked motion control system of wave maker.

The master station is a common PC with a standard Ethernet network card. And each slave station consists of network interface circuit board, 16-bit PIC Microcontroller, A/D and amplifier-filter circuit, servo driver interface circuit, multiplexer circuit for sensors, servo driver, and servo motor. In the following, we will present in detail the system with master station and slave station.

Circuits Principle of Slave Station

Figure 7 shows the designed circuit diagram of system. We will present the circuit principle in terms of the aforementioned four parts.

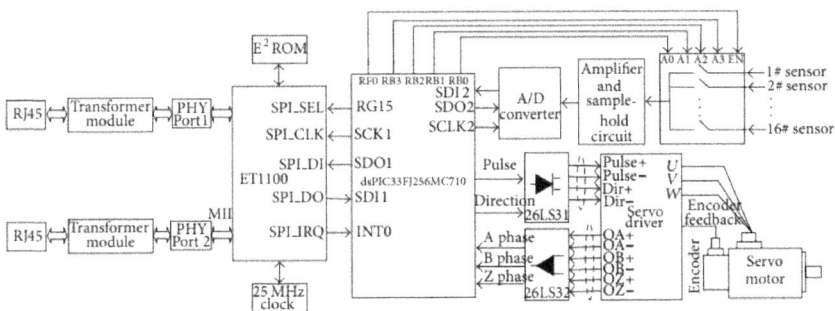

Figure 7: Structure of slave station module circuit.

(A) EtherCAT Network Interface Circuit [14]. In the circuit, the ASIC named ET1100 is regarded as EtherCAT slave controller, which serves as the Data Link Layer. The layer corresponds to Layer 2 in ISO model and provides real-time communication assurance among devices connected via EtherCAT network. Furthermore, the layer carries out the function of the frame check and accomplishes data transmission by extracting data from and/or inserting data into the Ethernet frame, based on the Data Link Layer parameters stored at predefined memory locations. However, the data frames transmission depends on network Physical Layer known as PHY. The layer can receive data bits stream from Data Link Layer and encode the bits into signals. Sequentially, the signals are sent to the transmission medium and received by the next Physical Layer interface. After being decoded, the signals are passed on to the Data Link Layer of the next slave controller. To perform the function of Physical Layer, KS8721BL, the Physical Layer interface chip is linked to ET1100 via Medium Independent Interface (MII). At the same time, the chip is connected to medium port connector RJ45 via network transformer HR601680 to improve the signal anti-interference capability. HR601680 is a 1 : 1 transformer with a smaller package and supports 10 M/100 M Ethernet. In addition to the circuit and parts mentioned above, ET1100 provides Serial Peripheral Interface (SPI) to link application layer controller. As an individual network interface board, it provides the 52 pins plug to connect with application controller. Via the plug, the ready data from Data Link Layer will be offered to the application layer microcontroller. The board contains the affiliated circuits, such as clock circuit which generates clock signals by a 25 MHZ crystal oscillator, EEROM circuit used to store device configuration parameters and device description information, reset circuit, and so forth.

(B) Application Controller Circuit. The application controller is the control core, and it takes charge of running the EtherCAT protocol program and application data accessing program. Due to the advanced motor control peripheral features, fast and efficient CPU, and small cost-effective package sizes, we adopt dsPIC Digital Signal Controller (dsPIC33FJ256MC710) as the core controller. The chip has up to 85 programmable digital I/O pins. In addition, several peripheral features are available including four timer/counters and eight Capture/Compare/PWM modules. Especially, there are two SPI modules in the chip, connecting the network interface controller and the network interface chip (ET1100), the network interface controller, and serial ADC, respectively. And the chip provides the control and data lines for all other circuits. It is a bridge linking the EtherCAT network interface circuits and field data acquisition and servo motor control circuits.

(C) Wave Height Data Acquisition Circuit. The data acquisition circuit of each slave station in the system, with the capacity of 16 simple analog channels input, is designed. Sixteen wave height sensors are connected with the channels input ports. Because the output signal from wave sensor is weak, only 0.2 V to 0.8 V, the collected signals must be amplified. To simplify the circuit design, the 16 sensor channels are selected by a multiplexer through the 4-bit binary address lines A0, A1, A2, and A3. These pins receive the address signals from the application microcontroller output pins RB3, RB2, RB1, and RB0 (see Figure 7). The channel number sent by the master station determines the values of the pins. According to the values, the multiplexer selects one of the collected 16-channel signals and forwards the selected input signals into a single outline. The output signal feeds into the amplifier which is capable of implementing gains 5. Hence the signal is amplified up to 1 V to 4 V. Finally, the analogue signals will be converted into digital signals via ADC. However, the ADC must have a stable signal to achieve conversion. Thus, the amplifier out signal is imported into a sample and hold circuitry before it is provided to ADC. The chip AD7276 is chosen in our system. It is 12-bit, high speed, low power, successive approximation ADC, which operates from a single 2.35 V to 3.6 V power supply and feature throughput rates of up to 3 MSPS (Million Samples per Second) and provides a SPI to connect with microcontroller.

(D) Servo Motor Interface and Control Circuit [15]. The part of circuit includes mainly pulse sending/receiving interface, servo motor encoder feedback interface, servo driver, and servo motor. The microcontroller receives the control commands from the master station and generates the signals of direction and pulse number. The servo motor accordingly rotates in a given angle. Pulses as a square wave are sent to servo driver via interface circuit, the number of pulses determines the angle of rotation, and frequency of square wave determines the speed of rotation. And then the driver controls the motor to move to destination position, which takes both speed and position feedback signals from the encoder of servo motor. The feedback values can be detected and calculated by the microcontroller. In Figure 7, the Minas A4 series driver and servo motor of Panasonic Company are selected and the driver is configured for position control mode. Furthermore, the motor is settled to take 10000 pulses to rotate 360 degrees. 27.8 pulses are required for a rotation of 1 degree. However, the fraction of pulse value can be omitted. So the motor is made to rotate in steps of 1.8 degrees which requires 50 pulses $((10000*1.8)/360 = 50)$.

State Machine and Object Dictionary's Definition

State machine is a very important concept in EtherCAT network communication. It is a series state transition responsible for the coordination of master station and slave station application at start-up and during operation. It is achieved both in master station and slave program. All of the state changes are initiated by the master station and the slave stations respond to the changes by executing corresponding program. The requested state by master station is written into the application layer control register and the response resulted by slave station is reflected in the application layer status register. The two registers are located in the chip ET1100. The detailed state transition diagram is shown in Figure 8.

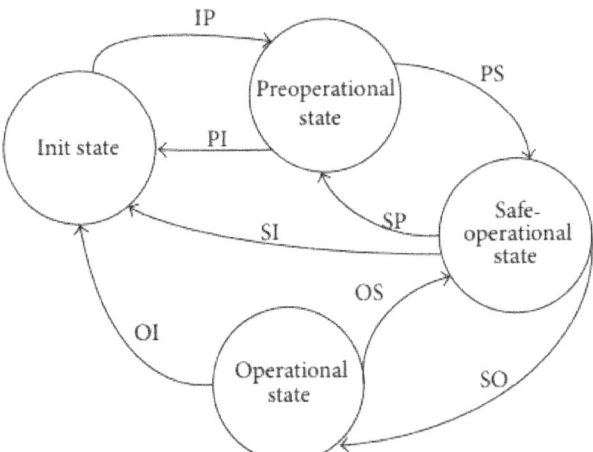

Figure 8: State transferring diagram.

The four states include [22] the following:

(i)   Init: the master station initializes the configuration registers of the chip ET1100 and configures the SMs for mailbox;

(ii)   pre-operation: the mailbox communication is activated;

(iii)   safe-operation: the mailbox and Process Data input communication is implemented;

(iv)   operation: both process data input and process data output are implemented.

Besides, the master will initialize the corresponding resisters of the chip ET1100 during state changes.

Object dictionary is another important concept in EtherCAT network communication. The EtherCAT protocol extends the object dictionary functionality of CAN bus standard. It defines all related data objects of the device in a standardized way. The object dictionary is made up of two sections. The first section includes general device information such as device identification, manufacturer name, and communication parameters. The specific device is functionally described in the second section. All of the accessed information is named as objects. According to the profile definition, a series of index number are used to describe the PDOs. In EtherCAT protocol, there are two types of PDOs including Sync Manager Channel objects and application objects. The indices 0x1C10 to 0x1C2F represent the Sync Manager Channel objects which describe a consistent memory area and manage several PDOs. The application objects are located at indices 0x1600 to 0x16FF for Receive PDOs (RxPDOs) and at indices 0x1A00 to 0x1AFF for Transmit PDOs (TxPDOs). The objects are read through the corresponding entries in the object dictionary. The object dictionary can be described by XML and it is downloaded into the EEROM on the network interface board using configuration tool. In order to access the PDOs rightly, the PDOs mapping relationship is represented in Table 2.

Table 2: PDOs mapping relationship

Sync channel 2 indices (process data output)		0X1C12
RXPDO indices	0X1600	0X1601
Input entry indexes	0X7000, 0X7001, ..., 0X70FF	0X7010, 0X7010, 0X7012
Input entry names	1# sensor, 2# sensor, ..., 16# sensor	encoder_A, encoder_B, encoder_Z
Sync channel 3 indices (process data input)		0X1C13
TXPDO indexes	0X1A00	0X1A01
Output entry indices	0X6000, 0X6001	0X6010
Output entry names	Motor_pulse, motor_direction	Channel_selection

Master Station Principle

As we can see in Figure 9, the master station comprises three main modules.

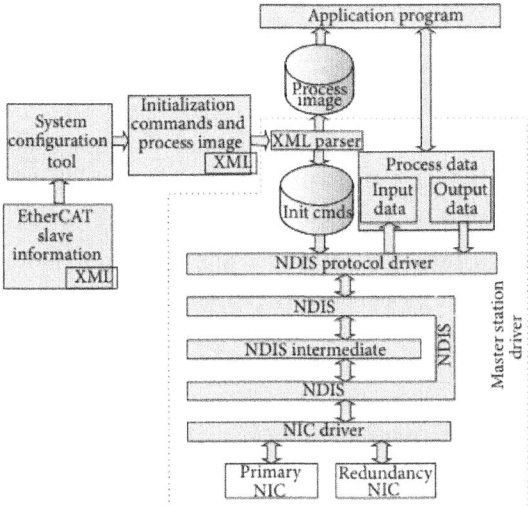

Figure 9: Describing for principle of master station.

The master driver module is the base of implementing EtherCAT communication. It includes a complete protocol stack with the following three types of drivers:

(i)   NIC Miniport Driver: this driver program is responsible for sending/receiving frames to/from NIC and provides an interface to upper protocol driver program;

(ii)   NDIS Intermediate Driver: the driver between NIC Miniport driver and Protocol Driver can control all traffic being accepted by the NIC. In order to implement EtherCAT protocol, the driver filters all Ethernet data frames and blocks all the frames except EtherCAT frames;

(iii)   NDIS Driver: it provides services for application layer clients [16].

Moreover, the master station driver can parse the XML which defines the initialization EtherCAT commands that should be sent during a specific transition according to the EtherCAT state machine and PDOs. And it is loaded through system configuration tool and is read from slave station EEROM during start-up by master station driver. The application program variables build a kind of mapping relationship with PDOs through XML definition. Our aim is to generate the wave height form based on the designed system accurately, the application program can implement the function of calculating and sending revised wave making pulse signals according to the given target wave sequence and the feedback data from field sensors and servo driver encoder. Therefore, a series of array names, such as motor_pulse, motor_direction, channel_number, encoder A, encoder_B, encoder_Z, and fb_sensor, are defined. Each defined array except the array fb_sensor possesses 10 member variables because of the system with 10 slave stations. There are 16 sensors at each slave station, so the total 160 sensors are linked to the system. Therefore, we must define the array fb_sensor including 160 member variables. All of the mentioned member variables are mapped with PDOs in the application program.

The application program sends the pulse signals and rotating direction signals to servo drivers by equal interval time. The target wave curve is converted into a series of discrete data dots. The number of pulse signals represents the data dots size. After the program sends the first group of signals, the feedback signals from servo drivers and wave height sensors are received. Then the next signals which will be sent are modified by the specific algorithm. The system goal is to make precise wave form in the experiment pool, which simulates the target curve wave form. The algorithm is described as "Algorithm 1".

```
    (i) Initialize PDOs image
Call Initialization ();
    {
    int motor_pulse[10] ← 0, motor_direction[10] ← 0;
    int encoder_A[10] ← 0, encoder_B[10] ← 0, encoder_Z[10] ← 0;
    int fb_sensor[160] ← 0, channel_number[160] ← 0;
    }
Execute LoadECATConfiguration ();
    {
    Load the *.xml file;
    /*brief Gets a pointer to the PDOs*/
    INT *ProcessDataPtr (INT imgId, INT inOut, INT offs, INT size);
    /*brief Gets the size of the PDOs*/
    INT ProcessDataSize (INT imgId, INT inOut);
    If (PDOs is output data)
    {
    UNSIGED LONG nData ← ProcessDataSize(0, VG_IN);
    INT *pData ← ProcessDataPtr(0, VG_IN, 0, nData);
    }
    Else if (PDOs is input data)
    {
    UNSIGED LONG nData ← ProcessDataSize(0, VG_OUT);
    INT *pData ← ProcessDataPtr(0, VG_OUT, 0, nData);
    }
    }
```

(ii) Send PDOs output data
motor_pulse[10] ← the sent pulse values of target curve to every channel;
motor_direction[10] ← the sent direction values of target curve to every channel;
channel_number[160] ← the selected feedback sensor numbers;
For (INT $i = 0, i < 10; i++$)
{
*pData = motor_pulse[i];
pData++;
*pData = motor_pulse[i]
pData++;
}
*pData ← channel_number[j], where $j = 0, \ldots, 160$;
pData++;
Sendpacket (pData, nData);
(iii) Receive PDOs input data
Receivepacket (pData, nData);
For (INT $i = 0, i < 10; i++$)
{
encoder_A[i] = *pData;
pData++;
encoder_B[i] = *pData;
pData++;
encoder_Z[i] = *pData;
pData++;
}
fb_sensor[j] ←*pData, where $j = 0, \ldots, 160$;
pData++;
Calculate the next output data according to the received data;
(iv) The program jumps the (ii).

Algorithm 1

PERFORMANCE ANALYSIS AND EXPERIMENT RESULTS

Based on the system features, the two most important performance indices including cycle time and delay jitter are analyzed and tested.

Definition of Cycle Time and Delay Jitter

The cycle time is defined as the necessary time to accomplish an input/output data exchange between the controller and all networked devices [7]. The cycle time T_{Cycle} can be calculated by

$$T_{Cycle} = T_{SM\_process} + T_{S\_frame} + T_{R\_frame}$$
$$+ T_{SP_{delay}} + T_{RP_{delay}} + T_{Idle} + T_{RM_{process}}.$$
(1)

In (1), $T_{SM\,process}$ and $T_{RM\,process}$ are the master sending processing time and master receiving processing time, respectively. They can be described by

$$T_{SM\_process} = T_{S\_AP\_process} + T_{S\_Pre\_frame} + T_{S\_PD\_process},$$

$$T_{RM\_process} = T_{R\_AP\_process} + T_{R\_Pre\_frame} + T_{R\_PD\_process}.$$

$$(2)$$

In (2), $T_{S\,AP\,process}$ is the time for application program calculating output variables and assigning them to the output $T_{S\_Pre\_frame}$ and $T_{R\_Pre\_frame}$ are the data frame queuing waiting time for sending and receiving data, respectively, and $T_{S\_PD\_process}$ and $T_{R\_PD\_process}$ are the protocol stack program processing time during which the data frame is sent and received.

In (1), T_{S_frame} and T_{R_frame} are the delay time of sending and receiving data frame, depending on the network communication bit rates; $T_{S\_Pre\_frame}$ and $T_{R\_Pre\_frame}$ delay represent the propagation delay time of sending and receiving the network data, including the wire propagation delay time, network controller forwarding delay time, slave microcontroller sampling, and processing delay time. T_{Idle} is the waiting time between the two sequentially sent frames. The time defined above can be shown in Figure 10.

Delay jitter in communications refers to the variations between the maximum cycle time and the minimum one [7]. Through this index we can decide whether the network is fast enough to handle the closed-loop servo system, or to use it only to download programs, send commands to the motion controller, and check the status of the controller for diagnostics and remote applications. Some systems can tolerate a little jitter. For example, data acquisition network communication systems that work with transmission rates of 100 milliseconds will not be affected by jitter. Likewise, distributed I/O handles cycle time of 10 milliseconds with little effect from jitter, while motion control systems work best with transmissions that contain less than 1 ms of jitter.

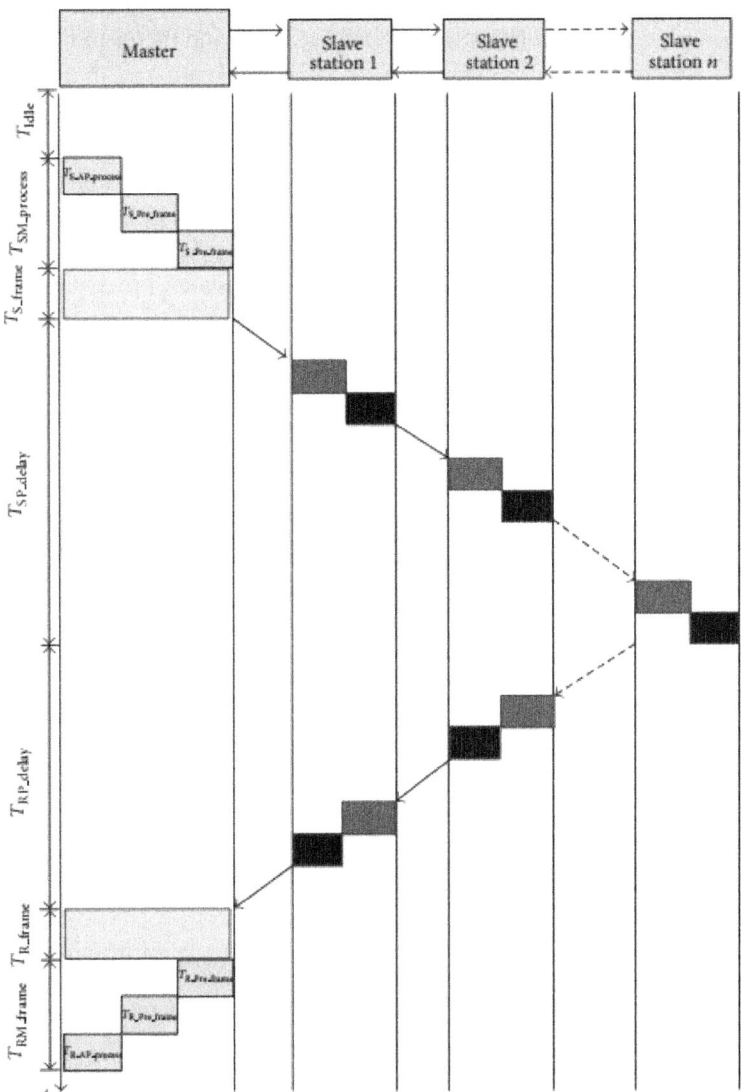

Figure 10: Network communicating cycle time.

Cycle Time and Delay Jitter of Different Protocols

The IEC 61784-2 standard defines a set of Performance Indicators (PIs) in order to specify the capabilities of the real-time Ethernet networks [28, 29]. As for the presented system in the paper, we focus on three main PIs: communication cycle data throughput and time synchronization accuracy. To verify the protocol performance superiority, we compare it with two representative Industry Ethernet protocols, namely, POWERLINK and PROFINET. The three protocols are comparable because they are standard Industry Ethernet protocol defined in IEC 61784-2 as a Publicly Available Specification. Moreover, the aforementioned two protocols have been widely used in the field of networked motion control. In particular, they achieve the whole master/slave-model cyclic real-time data exchange. But EtherCAT protocol exchanges the real-time data by using a so-called summation frame [17]. Because the comparison is more meaningful under the identical conditions, we will focus on a specific configuration, namely, the line topological architecture, comprising similar devices to the system designed in the paper. Furthermore, the length of data frame is ascertained as 128 bytes according to the system requirement and the network works at the speed of 100 Mbit/s. So the frame transfer time can be calculated as $T_{frame} = 10.24 \, \mu s$.

Moreover, we assume that the length of a copper-based segment between two devices is set to 50 meters, which leads to a medium delay of $T_{medium} = 227$ ns [17].

Cycle Time

In our analysis, the master process time is set as 200 μs. The communication cycle time can be written by

$$T_{Cycle} = T_{master\_process} + T_{S\_frame} + T_{R\_frame}$$

$$+ T_{SP\_delay} + T_{RP\_delay} + T_{Idle}, \quad (3)$$

where T_{S_frame} is equal to T_{R_frame}, and they can be calculated through the frame length and network speed. T_{SP_delay} is equal to T_{RP_delay}. Consequently, (3) can be defined as follow:

$$T_{Cycle} = T_{master\_process} + T_{tran\_delay} + T_{frame} + T_{Idle}, \quad (4)$$

where

$$T_{master\_process} = T_{SM\_process} + T_{RM\_process},$$

$$T_{tran\_delay} = T_{SP\_delay} + T_{RP\_delay},$$

$$T_{frame} = T_{S\_frame} + T_{R\_frame}. \quad (5)$$

According to the theoretical analysis provided in [17], $T_{\text{tran delay}}$ can be expressed as

$$T_{\text{tran\_delay}} = N_{\text{devices}} * \left(T_{\text{ecat\_fwd}} + T_{\text{medium}}\right),$$

$$T_{\text{frame}} = T_{\text{data}} + T_{\text{ethernet}} + T_{\text{ecat\_ov}}. \qquad (6)$$

We assume that there is no time delay between two frames ($T_{\text{Idle}} = 0$). Consequently, (4) can be defined as

$$T_{\text{Cycle}} = N_{\text{devices}} * \left(T_{\text{ecat\_fwd}} + 2 * T_{\text{medium}}\right)$$

$$+ T_{\text{frame}} + T_{\text{master\_process}}. \qquad (7)$$

Similarly, we get (8) from [16,17]. They represent the communication cycle time for POWERLINK and PROFINET protocol, respectively:

$$T_{\text{Cycle1}} = T_{\text{STA}} + N_{\text{devices}} * \left(T_{\text{SQ}} + T_{\text{QS}} + 2 * T_{\text{frame}}\right)$$

$$+ T_{\text{master\_process}},$$

$$T_{\text{Cycle2}} = N_{\text{devices}}$$

$$* \left(T_{\text{medium}} + T_{\text{pn\_fwd}} + T_{\text{frame}}\right) + T_{\text{master process}}. \qquad (8)$$

The other typical index values in (7) and (8) are given as follows:

$T_{\text{ecat fwd}}$: the forwarding delay time whose value is 2.7 μs;

T_{STA}: the duration of the start period with typical value 45 μs;

T_{SQ}: the time elapsed between the reception of a poll responses frame by the MN and the instant the Poll Requests frame is issued to the next CN and defined as 1 μs [16];

T_{QS}: the time employed by a CN to send a poll responses frame after the arrival of the Poll Requests from the MN with value 8 μs [16];

$T_{\text{pn fwd}}$: the forwarding delay time which value is 3 μs [17].

Figure 11 shows a diagram, whose compares the cycle time of EtherCAT, POWERLINK, and PROFINET as a function of the device amount. The diagram presents that the cycle time for EtherCAT protocol is smaller. And as the devices increase, the EtherCAT network cycle time increment is the least. When the devices increase from 1 to 10, the increment for EtherCAT protocol is only 28.75 μs. However, the incremental values are 265.32 μs and 121.2 μs for POWERLINK and PROFINET protocol, respectively.

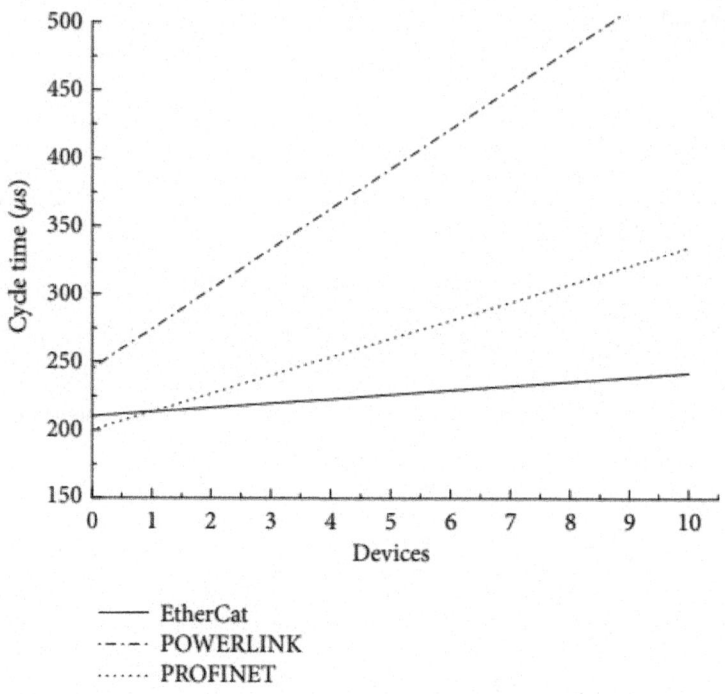

Figure 11: Communication cycle time of Three industry Ethernet protocols.

Data Throughput and Time Synchronization Accuracy

For this kind of network motion control system of ocean wave maker, all of the wave making boards often need to be simultaneously impeled. In other words, it would be better to send all of control commands and data in one frame by master station. And then it is received and executed by all of the slave stations at the same time, which need network protocol with high throughput and time synchronization accuracy. The data exchange method of EtherCAT protocol, using the summation frame, can access a large amount of slave devices over one standard Ethernet frame. So the data throughput can be up to 90% [26]. Considering the minimum cycle time, the reference [28] gives the throughput comparison results between EtherCAT and POWERLINK. It is obvious that the EtherCAT protocol is superior to POWERLINK protocol.

Furthermore, time synchronization accuracy can be up to 15 ns for the EtherCAT. The measured value by oscillograph is given in [30] whereas time synchronization accuracy of the other two protocols reaches only 1 μs [31, 32].

Experiment Results

We have tested the constructed networked wave maker control system which consists of 10 slave stations in three aspects. We adopt the system making regular sinusoidal wave and irregular wave, respectively, in the experiment shallow water pool. A series of regular sinusoidal data and irregular wave data are sent by the upper master station application program. The regular wave is a sinusoidal periodic signal with period 1 second and 6 cm wave height. And the irregular wave's averaged wave height is 6 cm. The wave height is known as the variance between the maximum value and the minimum one of the wave height curve in wave theory.

Figure 12 depicts the typical wave sensor feedback signals collected by the master station from 1# sensor, 16# sensor, 65# sensor, 145# sensor, and 160# sensor. As we can see in Figure 6, these sensors are linked to the 1# slave station, 5# slave station, and 10# slave station, respectively. Similarly, we make the irregular wave and the results are shown in Figure 13.

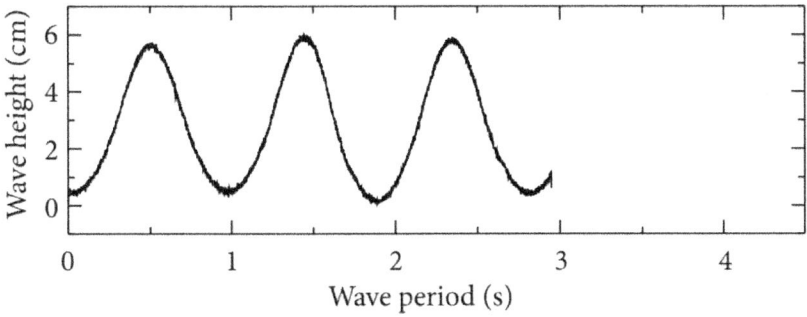

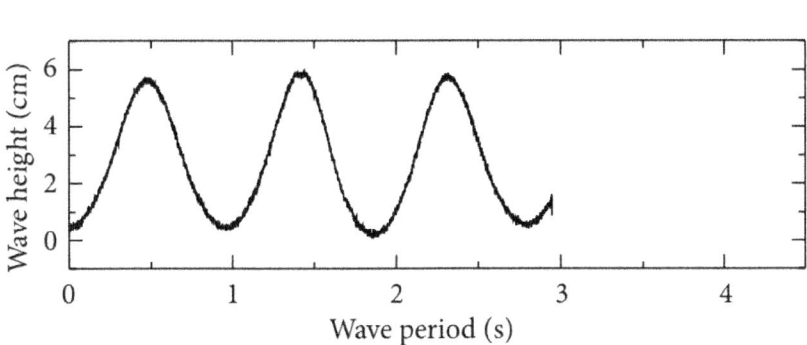

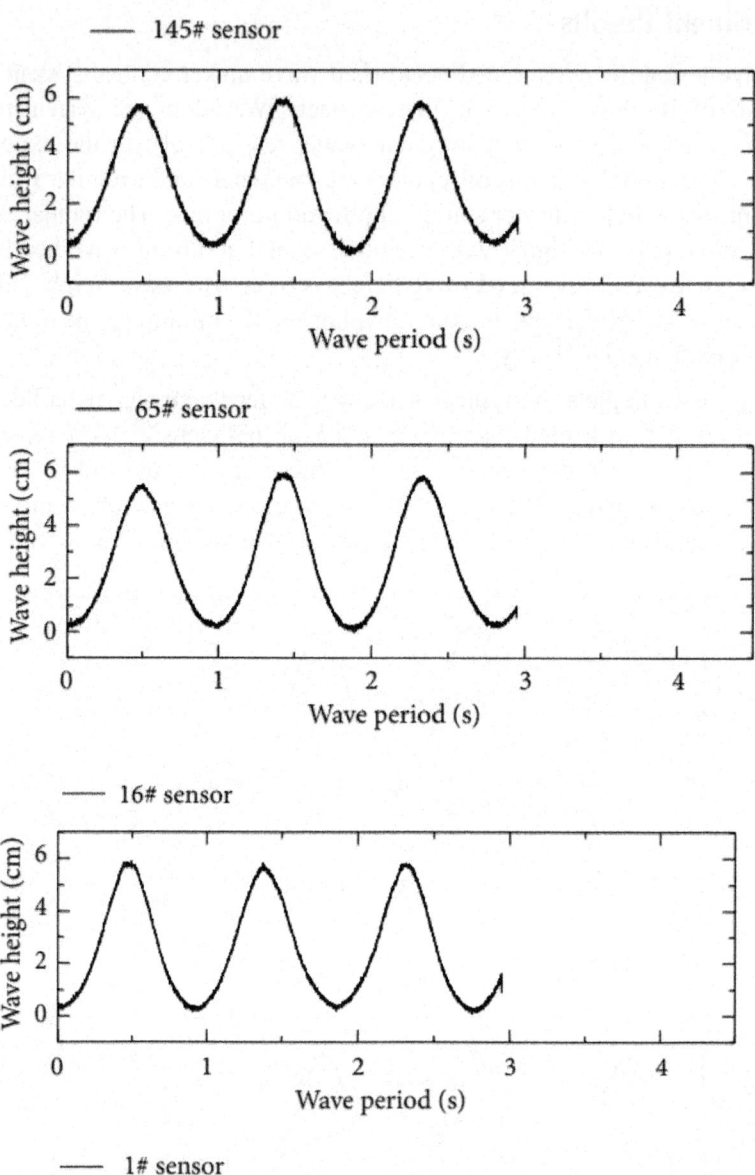

Figure 12: Regular wave height data signals from field sensors.

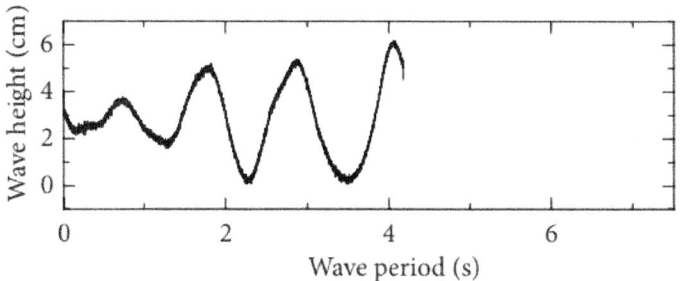

—— 160# sensor

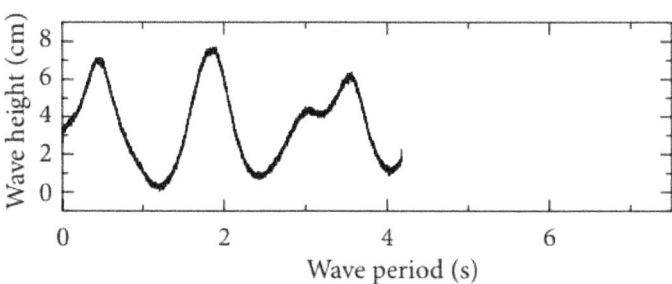

—— 145# sensor

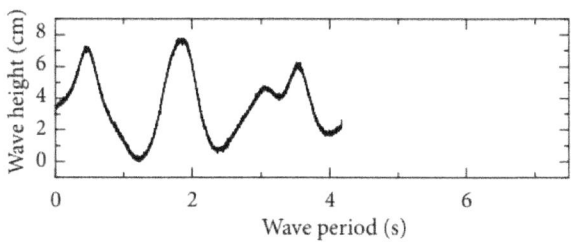

—— 65# sensor

—— 16# sensor

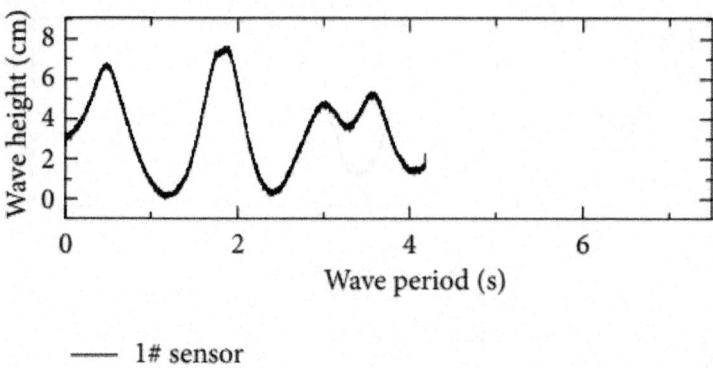

Figure 13: Irregular wave height data signals from field sensors.

The two diagrams of Figure 12 and Figure 13 show that the collected signal curves are smooth enough. Therefore, the system can meet making wave experiment needs.

Next, in order to validate the system communication performance, we have written program code to measure the cycle time. These codes implement the cycle time calculating and are inserted into the main application program. After the data frame with acquired data and servo control commands is sent by the master station, the timer starts until the data frame with the collected sensor data comes back to the master station. In order to illustrate clearly the advantage of EtherCAT, we tested the system with varying number of slave stations. The results are shown in Figure14.

The results show that the system including a different number of slave stations has almost the same average communication cycle time. But the average communication cycle time concentrates on about 250 µs. The delay time is very small when the data frame passes through the slave station node. The results agree with the theoretical analysis in Section 4.2. The value can be omitted in our system.

Moreover, we analyzed 10 groups of cycle time values and calculated the average values.

These values are depicted in Figure 15, which indicate that the maximum delay jitter is only 3.42 microseconds. The communication system can meet high-speed motion control requirement.

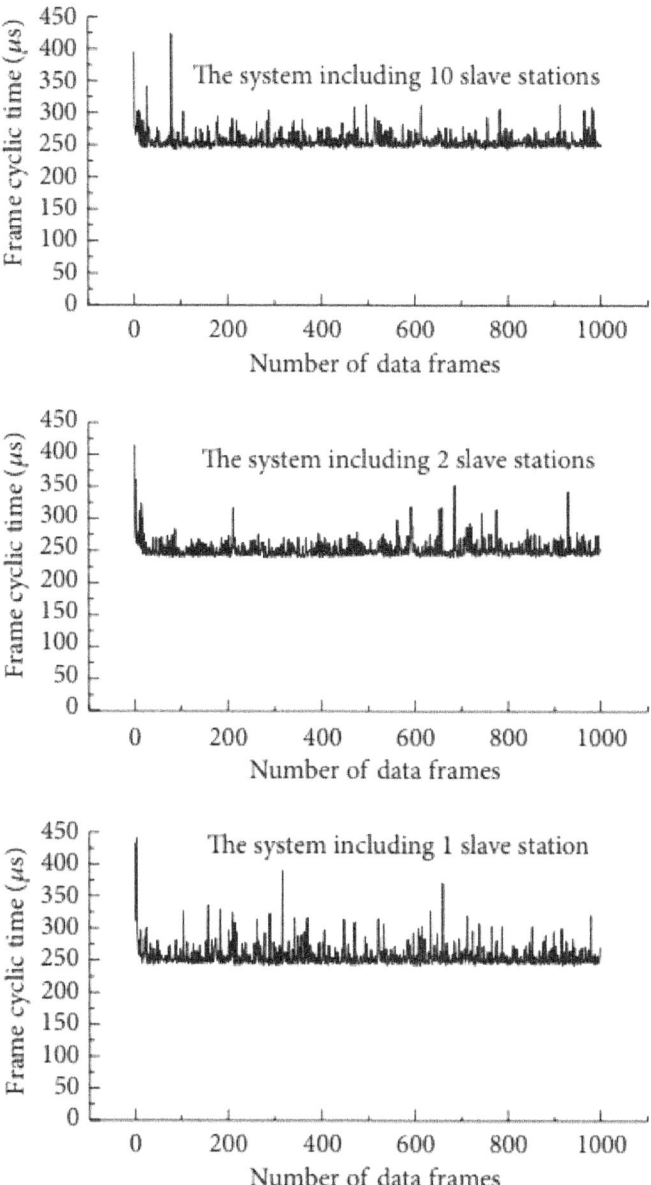

Figure 14: Cycle time of the system connected 1 slave station, 2 slave stations, and 10 slave stations, respectively.

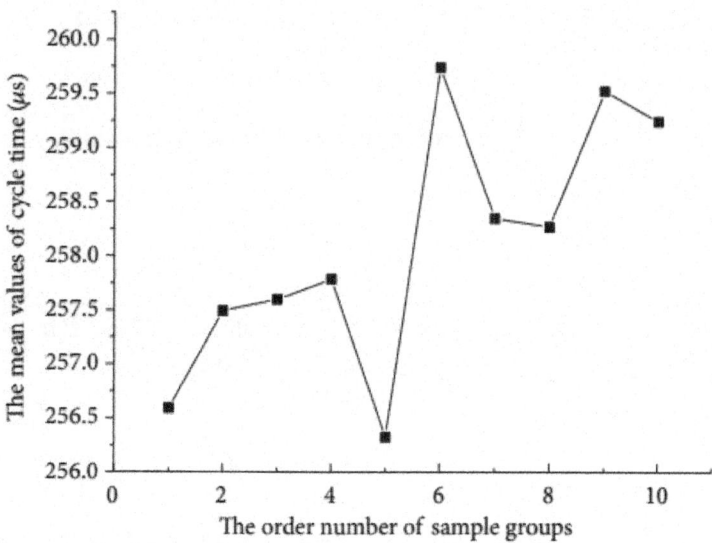

Figure 15: Average cycle time of system with 10 slave stations.

CONCLUSIONS

The paper introduced the design of a typical networked motion control system based on EtherCAT protocol. This networked motion control system with large amount of measurement sensors asks for not only achieving real-time motion control but also transferring numerous sensor signals by less than 1 millisecond communication cycle time. So far, dozens of Industry Ethernet protocols have been released. Different from other Industry Ethernet protocols, EtherCAT protocol possesses high communication speed and efficiency due to the specific communication principle. In particular, unparalleled hardware structure of slave interface controller makes the communication delay up to less than dozens of microseconds. Based on the test by control system platform, the experimental results show that EtherCAT protocol is the most suitable communication protocol for the type of networked motion control system mentioned. The results of this study would be helpful for the design of the networked motion control system with large-capacity data acquisition which requires high efficiency and high real-time.

CONFLICT OF INTERESTS

The authors declare that there is no conflict of interests regarding the publication of this paper.

ACKNOWLEDGMENTS

The paper is supported by Educational Commission of Henan Province of China (no. 2010A520020), National Natural Science Foundation of China (no. 51174263), Key Science and Technology Project of Henan Province (no. 112102210004), and Ph.D. Programs Foundation of Ministry of Education of China (20124116120004).

REFERENCES

1. K. C. Lee, S. Lee, and H. H. Lee, "Implementation and PID tuning of network-based control systems via Profibus polling network," Computer Standards and Interfaces, vol. 26, no. 3, pp. 229–240, 2004. · ·

2. N. P. Mahalik and A. N. Nambiar, "Trends in food packaging and manufacturing systems and technology," Trends in Food Science and Technology, vol. 21, no. 3, pp. 117–128, 2010.

3. Erturk, "A new method for transferring CAN messages using wireless ATM," Journal of Network and Computer Applications, vol. 28, no. 1, pp. 45–56, 2005.

4. Flammini, P. Ferrari, E. Sisinni, D. Marioli, and A. Taroni, "Sensor interface: from field-bus to ethernet and internet," Sensors and Actuators A, vol. 101, no. 1-2, pp. 194–202, 2002.

5. Abarca, M. de la Fuente, J. M. Abril, A. García, and F. Pérez-Ocón, "Intelligent sensor for tracking and monitoring of blood temperature and hemoderivatives used for transfusions," Sensors and Actuators A, vol. 152, no. 2, pp. 241–247, 2009.

6. Flammini, P. Ferrari, D. Marioli, E. Sisinni, and A. Taroni, "Wired and wireless sensor networks for industrial applications," Microelectronics Journal, vol. 40, no. 9, pp. 1322–1336, 2009.

7. J. Jasperneite, M. Schumacher, and K. Weber, "Limits of increasing the performance of industrial ethernet protocols," in Proceedings of the 12th IEEE International Conference on Emerging Technologies and Factory Automation (ETFA '07), pp. 17–24, Institute of Electrical and Electronics Engineers, Patras, Greece, September 2007. · ·

8. Tao, H. Ding, and Y. L. Xiong, "Design and implementation of an embedded IP sensor for distributed networking sensing," Sensors and Actuators A, vol. 119, no. 2, pp. 567–575, 2005.

9. S. Pal and A. Rakshit, "Development of network capable smart transducer interface for traditional sensors and actuators," Sensors and Actuators A, vol. 112, no. 2-3, pp. 381–387, 2004. · ·

10. L. Bissi, P. Placidi, A. Scorzoni, I. Elmi, and S. Zampolli, "Environmental monitoring system compliant with the IEEE 1451 standard and featuring a simplified transducer interface," Sensors and Actuators A, vol. 137, no. 1, pp. 175–184, 2007.

11. J.-D. Kim, J.-H. Lee, Y.-K. Ham, C.-H. Hong, B.-W. Min, and S.-G. Lee, "Sensor-Ball system based on IEEE 1451 for monitoring the condition of power transmission lines," Sensors and Actuators A, vol. 154, no. 1, pp. 157–168, 2009.

12. R. Dorschner, "Digital servo drives and SERCOS simplify machine installation, optimize performance," Control Solutions, vol. 73, no. 10, pp. 50–54, 2000.

13. "Ethernet-driven motion control goes on a real bender," Process and Control Engineering, vol. 59, pp. 21–23, 2006.

14. H. Makete, "Real-time requirements for discrete time applications in automation systems," Elektron, vol. 22, no. 3, pp. 36–40, 2005.

15. P. Ferrari, A. Flammini, and S. Vitturi, "Performance analysis of PROFINET networks," Computer Standards and Interfaces, vol. 28, no. 4, pp. 369–385, 2006.

16. G. Cena, L. Seno, A. Valenzano, and S. Vitturi, "Performance analysis of Ethernet Powerlink networks for distributed control and automation systems," Computer Standards and Interfaces, vol. 31, no. 3, pp. 566–572, 2009.

17. J. Jasperneite, M. Schumacher, and K. Weber, "Limits of increasing the performance of industrial ethernet protocols," in Proceedings of the 12th IEEE International Conference on Emerging Technologies and Factory Automation (ETFA '07), pp. 17–24, September 2007.

18. L. Seno and S. Vitturi, "A simulation study of Ethernet powerlink networks," in Proceedings of the 12th IEEE International Conference on Emerging Technologies and Factory Automation (ETFA '07), pp. 740–743, September 2007.

19. K. W. Song and G. S. Choi, "Fieldbus based distributed servo control using LonWorks/IP gateway/web servers," Mechatronics, vol. 20, no. 3, pp. 415–423, 2010.

20. L. Wang, P. Orban, A. Cunningham, and S. Lang, "Remote real-time CNC machining for web-based manufacturing," Robotics and Computer-Integrated Manufacturing, vol. 20, no. 6, pp. 563–571, 2004.

21. M. Rostan, J. E. Stubbs, and D. Dzilno, "EtherCAT enabled advanced control architecture," in Proceedings of the IEEE/SEMI Advanced

Semiconductor Manufacturing Conference (ASMC '10), pp. 39–44, July 2010.

22. A. GmbH, "Hardware Data Sheet ET1100—EtherCAT Slave Controller, Vel'. 1. 8," 2008, http://www.beckhoff.com.

23. G. Prytz and J. Skaalvik, "Redundant and synchronized EtherCAT network," in Proceedings of the 5th International Symposium on Industrial Embedded Systems (SIES '10), pp. 201–204, IEEE Computer Society, Trento, Italy, July 2010.

24. M. Rehnman and T. Gentzell, "Synchronization in a force measurement system using EtherCAT," in Proceedings of the 13th IEEE International Conference on Emerging Technologies and Factory Automation (ETFA '08), pp. 1023–1030, Institute of Electrical and Electronics Engineers, Hamburg, Germany, September 2008.

25. G. Cena, I. C. Bertolotti, S. Scanzio, A. Valenzano, and C. Zunino, "On the accuracy of the distributed clock mechanism in EtherCAT," in Proceedings of the IEEE International Workshop on Factory Communication Systems (WFCS '10), pp. 43–52, Institute of Electrical and Electronics Engineers, Nancy, France, May 2010.

26. G. Cena, S. Scanzio, A. Valenzano, and C. Zunino, "Performance evaluation of the EtherCAT distributed clock algorithm," in Proceedings of the IEEE International Symposium on Industrial Electronics (ISIE '10), pp. 3398–3403, Institute of Electrical and Electronics Engineers, Bari, Italy, July 2010.

27. J. C. Lee, S. J. Cho, Y. H. Jeon, and J. W. Jeon, "Dynamic drift compensation for the distributed clock in EtherCAT," in Proceedings of the IEEE International Conference on Robotics and Biomimetics (ROBIO '09), pp. 1872–1876, IEEE Computer Society, Guilin, China, December 2009.

28. L. Seno, S. Vitturi, and C. Zunino, "Real time ethernet networks evaluation using performance indicators," in Proceedings of the IEEE Conference on Emerging Technologies and Factory Automation (ETFA '09), September 2009.·

29. L. Winkel, "Real-time ethernet in IEC 61784-2 and IEC 61158 series," inProceedings of the IEEE International Conference on Industrial Informatics (INDIN '06), pp. 246–250, Institute of Electrical and Electronics Engineers, Singapore, August 2006.

30. EtherCAT Technology Group, http://www.ethercat.org/.

31. "Real-Time PROFINET IRT," http://www.profibus.com/.

32. Ethernet   Powerlink   Standardization   Group,   http://www.ethernet-powerlink.org/.

Chapter 2

OPEN-SOURCE HARDWARE IS A LOW-COST ALTERNATIVE FOR SCIENTIFIC INSTRUMENTATION AND RESEARCH

Daniel K. Fisher[1], Peter J. Gould[2]

[1]USDA Agricultural Research Service, Stoneville, USA

[2]US Forest Service, Pacific Northwest Research Station, Olympia, USA

ABSTRACT

Scientific research requires the collection of data in order to study, monitor, analyze, describe, or understand a particular process or event. Data collection efforts are often a compromise: manual measurements can be time-consuming and labor-intensive, resulting in data being collected at a low frequency, while automating the data-collection process can reduce labor requirements and increase the frequency of measurements, but at the cost of added expense of electronic data-collecting instrumentation. Rapid advances in electronic technologies have resulted in a variety of new and inexpensive sensing, monitoring, and control capabilities which offer opportunities for implementation in agricultural and natural-resource research applications. An Open Source Hardware project called Arduino consists of a programmable microcontroller development platform, expansion capability through add-on boards, and a programming development environment for creating custom microcontroller software. All circuit-board and electronic component specifications, as well as the programming software, are open-source and freely available for anyone to use or modify. Inexpensive sensors and the Arduino development platform were used to develop several inexpensive, automated sensing and datalogging systems for use in agricultural and natural-resources related research projects. Systems were developed and implemented to monitor soil-moisture status of field crops for irrigation scheduling and crop-water use studies, to measure daily evaporation-pan water levels for quantifying evaporative demand, and to monitor environmental parameters under forested conditions. These studies demonstrate the usefulness of automated measurements, and offer guidance for other researchers in developing inexpensive sensing and monitoring systems to further their research.

INTRODUCTION

Scientific research requires the collection of data in order to study, monitor, analyze, describe, or understand a particular process or event. Data collection efforts are often a compromise, however, between the amount and type of measurements needed and the resources available to collect them. Manual measurements can be time-consuming and labor-intensive, resulting in data being collected at a low frequency, with long time intervals between measurements. If outdoor field research is involved, collection intervals can be irregular when labor is unavailable, on weekends or when other duties take priority for example, or when inclement weather does not permit visits to the field. Automating the data-collection process can reduce labor requirements and greatly increase the frequency and regularity of measurements, but at the cost of added expense of electronic data-collecting instrumentation.

A vast number of electronic solutions are available for automated sensing, monitoring, and collecting information, but several problems exist which can limit their application in research work and acceptance by research scientists. Features, capabilities, and prices of commercially available datalogging instrumentation can vary greatly, from inexpensive, low-resolution, limited-input devices to expensive, full-featured, multi-input instruments. Developed by private industry, monitoring equipment often contains proprietary technology that manufacturers do not wish to release, and is often designed to operate with only a particular manufacturer's sensors. The user can become locked into a particular manufacturer's systems or sensor technology due to high costs of the monitoring equipment making it cost-prohibitive to switch to a different vendor. If a range of different sensor information is desired, a single vendor may not supply all that is needed, and several monitoring systems may be required due to incompatible technologies. Since the scientific data-collection and monitoring market is small, private companies may be slow to innovate or introduce new technologies based solely on economic analyses. And to obtain sufficient quantities of data from an experiment, multiple sites and replicated treatments may be needed to satisfy observational and statistical requirements, which can quickly become cost-prohibitive.

Rapid advances in electronic technologies have resulted in a variety of new and inexpensive sensing, monitoring, and control capabilities. These rapidly evolving technologies provide researchers and practitioners with access to low-cost, solid-state sensors and programmable microcontroller-based circuits.

Microcontrollers can be thought of as small, low-power, low-cost computers packaged within a single chip. The microcontroller runs a program that is created and uploaded by the user to operate different components within a circuit. The user can modify the program and change the function of the circuit without changing the circuit physically. Many types of sensors and auxiliary components, such as memory chips, clocks, and communications devices, are available which interface directly with microcontrollers, simplifying circuit designs and putting electronic design within reach of people with limited electronics background and knowledge. A number of microcontroller-based devices have been described in which the specific requirements of a research project dictated the development of customized monitoring systems with unique capabilities [1-5].

A further advancement in microcontroller-based sensing and monitoring relates not specifically to the design and development of the electronics and physical components, but to the idea of making the designs and development efforts freely available to all in order to facilitate and expand the adoption of the technologies. The rapid rise of the internet and accessibility of computer resources led to the concept of Open Source Software as a means to provide free and transparent access to computer code so that individuals could review, modify, improve, and distribute computer software (Open Source Initiative, http://www.opensource.org). In recent years, a similar effort was undertaken to enable the free and open sharing of hardware designs and projects so that, by sharing and collaborating with others who have similar interests and needs, innovation could occur more quickly, improvements could be suggested and incorporated, and more users could access the final product.

One such Open Source Hardware project resulted in the creation of a microcontroller-based development platform called Arduino [6]. The Arduino hardware consists of a programmable microcontroller mounted on a circuit board which provides convenient access to the microcontroller input/output pins and connectivity to a personal computer for programming and user interaction. The circuit board has a standardized size and physical configuration so that any Arduino-compatible boards can be interchanged. Standardized add-on boards (called shields) plug into the Arduino circuit board, and are used to expand the capabilities of the main board. The microcontroller is programmed via the Arduino Integrated Development Environment (IDE), in which the user creates the program instructions to operate the microcontroller and then downloads the program to the microcontroller. As an open-source hardware project, all circuitboard and electronic component specifications, as well as the IDE software, are freely available for anyone to use or modify. As a result, private manufacturers all around the world produce and offer inexpensive, standardized Arduino-compatible hardware with an extensive supply of features and capabilities. Researchers have begun to develop and implement devices based on the Arduino platform for a variety of applications [7-12], with ease of use, low cost, and standardized components and programming language cited as reasons for choosing the Arduino platform.

The objective of this paper is to introduce researchers and practitioners to potential applications of the opensource Arduino platform for implementation in research and monitoring applications. Specifically, we 1) describe the Arduino microcontroller development platform, 2) discuss examples of sensing and auxiliary circuit components available, and 3) demonstrate several datalogging devices developed for use in agricultural and natural-resources research.

COMPONENTS

Arduino Microcontroller Development Platform

The current standard Arduino development platform is based on an ATmega328 8-bit programmable microcontroller (Atmel Corporation, San Jose, CA USA). A printed-circuit board positions the microcontroller in a circuit so that the input/output (IO) pins are easily accessible. The microcontroller contains 32 kilobytes (KB) of flash memory for program storage and 1 KB of non-volatile data-storage memory. IO lines consist of 14 digital pins and 6 analog pins, which provide 6 channels of 10-bit analog-to-digital (A/D) conversion capability. The microcontroller contains many built-in features, including timer/counters, internal and external interrupts, serial and other communication-protocol capabilities, programmable watchdog timer, and low-power, energy-saving modes.

Versions of the Arduino board are available which use other, more-powerful microcontrollers, have additional IO pins, and have different physical sizes. Devices operate at either a 5-V level and oscillator speed of 16 MHz or a 3.3-V level and 8 MHz. While many boards have an on-board USB connector to interface with a personal computer, the ATmega microcontroller communicates via a two-wire serial (transmit, Tx, and receive, Rx) connection. Boards with on-board USB connector also have a USB-serial converter chip and use a standard USB-USB cable, while other boards, to simplify design and lower cost, do not incorporate the USB-serial chip. A special cable, which contains the USB-serial chip and creates a virtual serial port, must be used.

The Arduino board is designed to allow expansion through the connection of auxiliary boards or shields. The shields connect via mating pins which are arranged in the same physical configuration as the Arduino board, and simply plug onto the headers on the top of the Arduino board. The shields are then controlled by the Arduino microcontroller and program, which access the shields' pins through the Arduino pins. Programming libraries allow users to quickly integrate new devices and sensors into projects without needing to write extensive new program routines.

Software

The software environment for programming and interacting with the Arduino board is available for download and installation for several computer operating systems (GNU/Linux, Mac OS X, and Windows). Using the IDE, the user writes programs in a language based on C++. The IDE then compiles and error-checks the program, and downloads the compiled routine to the microcontroller. A terminal window is available for outputting text and data from the Arduino board to the computer monitor and for interacting with the microcontroller.

As an open-source project, the Arduino benefits from the collective efforts and expertise of developers from around the world. Programming libraries, which contain routines to simplify programming and incorporate advanced features, sample code, and complete programs are available to download, use, and modify as needed. The IDE, libraries, and sample code can be accessed via the Arduino project website [6].

Communications

The Arduino development platform provides several methods of communicating with external components, sensors, and computers. In addition to built-in A/D converters and timers for measuring analog voltage signals, several standardized communications protocols are available for interfacing digital components and sensors.

The Inter-Integrated Circuit, also called I²C or I2C, protocol developed by Philips Semiconductor, is a twowire serial transfer protocol designed for communications between integrated-circuit chips and microcontrollers. Two IO pins on the Arduino's ATmega328 microcontroller are designated for I2C communication. Each I2C device has its own unique identification number and address, allowing multiple devices to be connected to the same I2C pins. The microcontroller initiates communication with a device by first sending the address of the device and then reading data from or writing data to the device. Identification numbers are unique to each type of component (memory chip, clock, temperature sensor, etc.) while addresses are either preset by the manufacturer or specified by the user through different hardware configurations.

The Dallas 1-Wire protocol, developed by Dallas Semiconductor, uses a single IO pin for communication and, optionally, to power the external 1-Wire device. Like I2C, multiple devices can be connected to a single 1-Wire pin, and are called by the microcontroller using the device's unique address.

The Serial Peripheral Interface, or SPI, is a four-wire system developed by Motorola and provides a serial data link that operates in full duplex mode. SPI devices communicate in master/slave mode using three IO pins, with the master device, the microcontroller, initiating communications with the slave, a sensor or other device. The microcontroller uses an additional IO pin for each device to select and communicate with a particular device.

RS-232 is the standard serial communication protocol that was widely used to communicate between personal computers and peripherals before the advent of the universal serial bus (USB). RS-232 uses two communication lines (Rx to receive, Tx to transmit), and is the protocol used by the Arduino' microcontroller to interface with a computer for programming. Since few modern computers contain an RS-232 port, a virtual serial port must be created. While some Arduino boards have a USB-to-serial converter chip on-board, many boards do not in order to reduce cost and power consumption. A special USBserial cable which contains the converter chip, such as the FTDI Cable (www.makerspace.com), interfaces to the

Sensors

A large number of sensors are available to monitor and measure many types of environmental parameters or physical processes. The rapid advances and usage of programmable microcontrollers have brought an increase in the availability and ease of use of sensing devices designed to interface with microcontrollers. The sensors operate at low voltages, and output signals compatible with microcontrollers, including analog voltages, varying frequencies, and a selection of digital communications protocols.

While the number of parameters sensed, and the number of sensors available, is vast, a few examples are presented and discussed in the following subsections.

Temperature

One of the most-common measurements made in a multitude of disciplines is temperature. A variety of temperature sensors is available using several different measurement technologies. While thermistors, which are sensors whose electrical resistance changes in response to temperature, are still in use, alternate electronic sensors are available which are designed to interface easily with microcontrollers and computers.

Analog temperature sensors, such as the LM35 (National Semiconductor, Santa Clara, CA USA) and TMP36 (Analog Devices, Inc, Norwood, MA USA), are designed to output a voltage signal proportional to temperature. The microcontroller supplies an excitation voltage to the sensor, and then measures the sensor's output voltage with an on-board A/D converter. The microcontroller program calculates temperature using a calibration developed by the sensor manufacturer. The LM35 sensor, for example, provides a linear response with a calibration of 10 mV/C: temperature (°C), is therefore calculated by dividing the output voltage, in mV, by 10. Analog sensors are usually very inexpensive and easy to work with, requiring only a simple voltage measurement and calibration equation to determine temperature. The microcontroller must have an A/D converter, and a stable reference voltage, which some may not have, requiring the addition of external components and circuitry.

Digital temperature sensors are designed to provide a calibrated and voltage-converted output which can be read directly as a temperature value. These sensors do not require a voltage measurement to be made, allowing the use of microcontrollers which do not have A/D converters. Digital sensors interface with the microcontroller through one of several communications protocols, such as I2C, 1-Wire, and SPI, with transfer of information accomplished via the microcontroller program. Digital sensors often have the feature of a unique identification number, allowing multiple sensors to be connected to the same IO pins on the microcontroller, thus not using additional pins. In contrast, since each analog sensor would require its own A/D input pin, multiple analog sensors could quickly fill available A/D converter pins.

For making non-contact temperature measurements, infrared thermometer (IRT) sensors are available which are inexpensive and easy to interface. The MLX90614 (Melexis SA, Ieper, Belgium) series of IRTs communicate with the microcontroller via the I2C protocol. Experience using these sensors to monitor crop canopy temperature [5] has shown them to work well in a harsh agricultural environment, operate for extended periods under battery power, and provide accurate temperature measurements.

Soil-Water Status

In many agricultural, natural-resource, and water-management disciplines, water availability and moisture status are of great importance. The amount of water available in the soil profile for extraction by growing plants can be measured with a water-content sensor. A water potential sensor provides a measure of how tightly the water is held to the soil particles and how much energy must be expended to extract the water by the plant roots. This can be related to the availability of water to the plant.

Many of the currently available water-content sensors rely on a measure of the capacitance of the soil-water environment. Dielectric properties of the soil-water system vary weakly with soil properties, such as mineral composition, bulk density, and organic-matter content, but are strongly influenced by water content [13]. Watercontent sensors, such as the EC-5 and EC-20 (Decagon Devices, Pullman, WA USA), and VG400 (Vegetronix, Bluffdale, UT USA), consist of a capacitive-sensing element and on-board electronic circuitry. When powered by the microcontroller, the sensors return a voltage signal proportional to the water content in the soil. Measuring the voltage with the microcontroller's A/D converter and applying a calibration equation in the microcontroller program results in a water-content value, expressed in units of volume of water/volume of soil. Sensor manufacturers may provide calibration equations for limited soil types and other porous media, such as potting soil or greenhouse media, but the user often must develop a calibration, or at least verify the manufacturer's, under his specific soil conditions to obtain accurate water-content measurements.

Water-potential sensors are usually designed to act as variable resistors, in which the electrical resistance of the sensor varies in response to its water content. The sensor is composed of a porous matrix, and water can move into and out of the matrix in response to the matric potential of the soil. As the water content in the porous matrix changes with matric potential, the electrical resistance also changes. A calibration equation then converts resistance to matric or water potential, expressed in units of kiloPascals (kPa).

The Watermark 200SS (Irrometer Company, Riverside, CA USA) water-potential sensor is popular in irrigation-scheduling applications due to its ease of installation and low cost. It requires an alternating-current excitation rather than direct current, however, which can involve additional care when interfacing with a microcontroller (see [3] and the discussion in Section 3.1 below for alternative implementations). To allow direct connection and use with any microcontroller circuit, the MPS-2 (Decagon Devices, Pullman, WA USA) is designed to operate from a direct-current supply and output a simple voltage signal in response to soil-water potential. The voltage signal is measured with the microcontroller's A/D converter and then converted to water potential with a calibration equation.

Distance/Height

Distance measurements are common in robotic and industrial/manufacturing environments to determine distance from a moving vehicle for obstacle avoidance, detect presence or absence of material, and ensure proper placement of a component. In research applications, distance measurements can be used to determine properties such as plant height and canopy width, depth of water in canals, and fluid levels in tanks.

Distance measurements are commonly made using two sensing technologies, ultrasonic and infrared. Ultrasonic sensors often consist of two transducers, one which emits a pulse of high-frequency sound waves, and a second one to detect the sound after reflecting off a nearby surface. Distance is determined by measuring the length of time between sending the pulse and receiving the reflection, or echo, and converting this to a distance based on the speed of sound. Ultrasonic sensors, such as the SRF series (Devantech Ltd., Norfolk, UK) and the PING (Parallax Inc., Rocklin, CA USA) interface with a microcontroller via one or two digital IO pins. The microcontroller is programmed to initiate a pulse, then starts an internal timer and counts the number of microseconds until an echo signal is detected, and calculates the distance based on this time interval. Sensors are available with varying fields of view to enable sensing over wider or narrower regions.

Infrared sensors operate by emitting a beam of light and detecting the reflected beam, after hitting an obstacle, with a light sensor. The reflected beam returns at a slight angle from the emitted beam, and the angle of the two beams is dependent on the distance of the obstacle from the sensor. The reflected beam strikes the light sensor at some point, and is read by an on-board microcontroller which is programmed to output an analog voltage in proportion to distance. The analog voltage is input to the Arduino microcontroller's A/D converter and converted to distance with a calibration equation supplied by the manufacturer. Infrared sensors such as the GP2 series (Sharp Electronics Corporation, Mahwah, NJ USA) offer a variety of operating ranges.

Pressure

Maintaining proper pressure and measuring the existing pressure are important in many processes and environments. Atmospheric air pressure is an important meteorological parameter, for example, and liquid pressure can be used to determine fluid depth based on hydrostatic pressure relationships.

Many pressure sensing devices are available and range from simple sensing elements to amplified, calibrated, and temperature-compensated sensors. Sensing configurations typically consist of piezoresistive elements and a silicon diaphragm arranged in a Wheatstone-bridge circuit. A change in pressure causes the diaphragm to flex and changes the resistance values of the piezoresistive elements. Since changes are very small, the change in electrical output of the Wheatstone bridge is also small, requiring accurate voltage-measuring circuitry. Amplifying the output signal allows the signal to be measured with an A/D converter on the Arduino. Temperature changes can also affect the piezoresistive elements, resulting in the need for temperature compensation under conditions of large temperature swings. A range of pressure sensors, including the non-temperature-compensated 24PC, temperature-compensated 26PC, and fully compensated and amplified 40PC series (Honeywell Sensing and Control, Golden Valley, MN USA) can be interfaced and read with the Arduino's microcontroller.

Resolution of Analog Sensor Measurements

Analog sensors output a voltage signal which is converted into a numerical value by an A/D converter. The A/D converter is characterized by a known, reference voltage, which determines the range of acceptable voltage signals, and the number of digital values, or bits, into which the voltage range is divided. The Arduino's microcontroller contains a 10-bit A/D converter, meaning that the voltage range is divided into 2^{10}, or 1024, divisions. To measure a sensor's voltage signal, the A/D converter compares the voltage level to the reference voltage, and returns a proportional digital value in the range of 0 to 1023.

The A/D converter characteristics determine the resolution and accuracy of voltage measurements. The resolution, or smallest change in voltage that the A/D converter can detect, is dependent on the A/D converter's number of bits and the reference voltage. The Arduino's microcontroller has a built-in 1.1 V reference, which provides the A/D converter with a resolution of 1.1 V/ 1024 bits, or 0.00107 V/bit. The microcontroller's 5-V power supply voltage can also be used as a reference, resulting in an A/D conversion resolution of 0.00488 V/bit.

Resolution can be increased or decreased by changing the number of A/D conversion bits. External A/D converter chips are available which have higher-bit resolutions and can be easily interfaced with the Arduino. The MCP3424 (Microchip Technology Inc., Chandler, AZ USA) is an A/D converter chip which can read four input voltage signals with 18-bit (262,144 divisions) resolution. With a 5-V reference voltage, this would provide an A/D resolution of 0.0000191 V/bit. The MCP3424 communicates with the microcontroller using I2C.

To illustrate the effect of A/D converter resolution on sensor measurements, consider an analog temperature sensor that outputs a voltage signal between

0 and 5 V over a temperature range of 0 to 65 C. The signal, therefore, changes by 65 C/5 V, or 13 C/V. Using the microcontroller's built-in 10-bit A/D converter and a 5-V reference, with a resolution of 0.00488 V/bit, the resolution of temperature measurements would be 13 C/V*0.00488 V/bit, or 0.06 C/bit, which would be acceptable for most applications.

The resolution of a signal from a non-amplified pressure sensor, with an output of 0 to 10 mV over a range of 0 to 100 kPa, would have a measurement resolution of 100 kPa/0.01 V*0.00488 V/bit, or 48.8 kPa/bit. This would be unacceptable, providing only three measurements (0, 48.8, and 97.6 kPa) over the entire measurement range. Using the MCP3424 external A/D converter, with 18-bit resolution, would greatly improve voltagemeasurement capability and provide a pressure-measurement resolution of 0.038 kPa/bit.

Time-Keeping

In many data-collection efforts, proper timing of measurements and dateand time-stamping of sensor data are required. The microcontroller on the Arduino board has a very accurate 16 MHz oscillator and the ability to measure time increments with microsecond accuracy, but is not designed to provide real time (hours, minutes) and date information. If electrical power to the microcontroller is lost, the oscillator and microcontroller program cease to function, and any timing information is also lost.

External real-time clock (RTC) chips are used to provide time-keeping functions, with dedicated built-in or added backup batteries to retain accurate time information. RTCs such as the DS1307 and DS1337 (Maxim Integrated Products, Inc., Sunnyvale, CA USA) interface with the microcontroller using the I2C protocol, while others, such as the MCP795 (Microchip Technology Inc., Chandler, AZ USA) communicate via SPI. Simple routines in the microcontroller program access the RTCs to set or read time and date information, which can then be used to trigger sensor measurements at regular time intervals or record timing information of events.

Data Storage

Data collection often involves long-term, automated storage of sensor measurements. While the Arduino's microcontroller has extensive memory available for program storage, non-volatile data-storage capability is limited. On-board memory consists of 1 kb (1000 bytes), so a maximum of 1000 data values could be stored and retained if battery power were interrupted. To expand the storage capacity, external storage must be added.

External memory chips are available with varying amounts of non-volatile memory. The 24LC family of memory chips (Microchip Technology Inc., Chandler, AZ USA), for example, are available in capacities from 16 bytes to 65,356 KB. These chips communicate via the I2C protocol and have individual identification numbers so that multiple chips could be connected to increase storage amounts considerably.

For permanent or large-capacity storage, add-on boards are available which provide data storage to standard SD memory cards (Adafruit Industries, New York, NY USA) or microSD memory cards (Sparkfun Electronics, Boulder, CO USA). Memory cards are commonly available with storage capacities from 1 gigabyte (GB) to several GB, are inexpensive, and can be easily interfaced with the Arduino hardware. Since the memory cards can be read with a computer, data can be transferred quickly and easily between datalogger and computer. Software libraries have been written to provide all memory card reading, writing, and data-access functions, enabling rapid incorporation of memory-card storage into a datalogging project.

SENSING APPLICATIONS

To illustrate how the Arduino platform can be used to develop and implement an inexpensive, automated data collection and monitoring program, several examples are presented. These examples include a brief description of the circuitry and details of the project implementation. Microcontroller programs are not included but are freely available by contacting the authors.

Soil-Moisture Monitoring Datalogger

Monitoring moisture status of the soil profile is useful in scheduling irrigations and monitoring the movement or availability of water in the soil profile. Sensors are installed in the soil profile at various depths within a crop's root zone and are monitored periodically. A datalogger was designed to record measurements from three soilmoisture sensors at one-hour intervals, and store the measurements, along with the date and time, to a microSD memory card.

Hardware

The main components of the datalogger include an Arduino-compatible microcontroller board, voltage regulator, microSD/prototyping shield, and real-time clock/ calendar. The Diavolino microcontroller board (Evil Mad

Science LLC, Sunnyvale, CA USA) was chosen for its low cost, simple and low-component design, and ease of modification for battery-powered operation. The board, designed to operate from the 5 V power supplied via the USB computer connection, was modified by adding a two-pin header to connect an external AA battery pack. An LP2950 voltage regulator (National Semiconductor Corp., Santa Clara, CA USA) and capacitors were added to convert the unregulated battery voltage to a stable 5-V source to power the microcontroller. A trace on the printed circuit board, which powered the board from the USB connection, was then cut so that the only power source was the AA battery pack. The modified Diavolino microcontroller board is shown in **Figure 1**.

The microSD shield (Sparkfun Electronics, Boulder, CO USA) consists of a microSD-card holder, with on-board voltage-level shifter to supply the proper voltage levels for reading from and writing to a microSD card, and a prototyping area to incorporate additional circuitry into the shield. The microSD shield was designed to be powered from the microcontroller board's power supply, thus the microSD card and voltage-level shifter would always be powered and continuously drawing current. The shield was modified for battery-powered operation by rerouting the power supply for the microSD card and voltage-level shifter to one of the microcontroller's digital pins so that the components could be turned on and off as needed. A microSD card (Samsung) with a 2 gigabyte storage capacity was then inserted into the microSD card holder.

A circuit was designed and added to the microSD shield's prototyping area to measure the output from three soil-moisture sensors. A DS1337 real-time clock/ calendar chip provides date and time information for the microcontroller to make sensor readings at regular time intervals and to dateand time-stamp sensor data stored to the microSD card. A 32.768 kHz crystal oscillator provides an accurate timing signal for the DS1337, and a 3.3-V lithium coin cell battery powers the clock chip. The DS1337 interfaces with the microcontroller via the I2C protocol.

The soil-moisture sensors consist of three Watermark 200SS matric-potential sensors whose electrical resistance varies with moisture content. A circuit was designed in which each sensor, which acts as a variable resistor, forms one leg of a half bridge, or voltage divider.

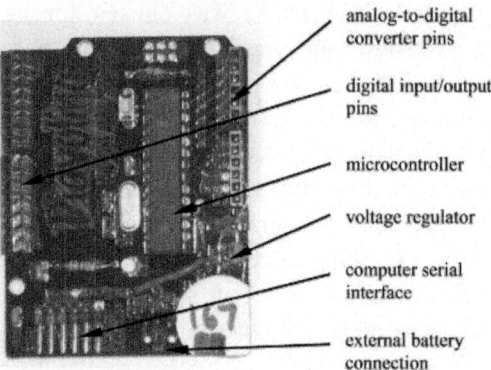

analog-to-digital
converter pins

digital input/output
pins

microcontroller

voltage regulator

computer serial
interface

external battery
connection

Figure 1: Modified Diavolino Arduino-compatible microcontroller board.

The half bridge is connected to two digital pins on the microcontroller, and each voltage-divider output is connected to an A/D pin. A photograph of the completed circuit, mounted on the microSD shield, is shown in **Figure 2**, and a schematic of the circuit is shown in Figure 3. A list of materials, with sources and approximate cost (small-quantity retail price, in the United States, US dollars, 2011), is provided in **Table 1**.

Software

Using the Arduino IDE installed on a personal computer, a microcontroller program, called a sketch on the Arduino platform, was written to read the real-time clock, make soil-moisture sensor measurements, and store the time and sensor data to a microSD card. Communication between the computer and microcontroller board requires an RS-232 serial connection, which was accomplished via an FTDI USB-serial cable, which interfaces to the computer's USB hub and creates a virtual serial port.

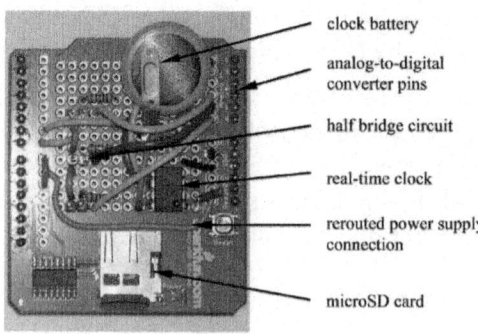

clock battery

analog-to-digital
converter pins

half bridge circuit

real-time clock

rerouted power supply
connection

microSD card

Figure 2: Modified microSD shield with circuit components installed.

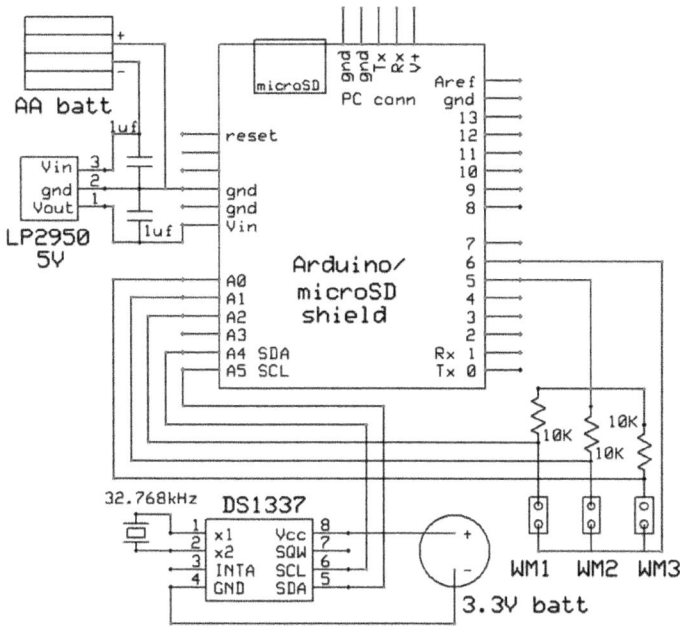

Figure 3: Schematic for soil-moisture sensor datalogger.

Table 1: List of materials for soil-moisture sensor datalogger

Main components	Part number	Supplier	Cost $
Microcontroller board	Diavolino	Evil Mad Science	13
microSD/prototyping shield	microSD shield	Sparkfun Electronics	14
microSD card	2 gb	Samsung	4
Real-time clock/calendar	DS1337	Maxim Integrated Products	3
Oscillator	32.768 kHz	Citizen America	1
Regulator	LP2950 5V	National Semiconductor	2
Miscellaneous (capacitors, resistors, headers, batteries)			7
Datalogger Total			44
Soil-moisture sensor	200SS	Irrometer	30

To enable long-term, battery-powered operation of the datalogger, the microcontroller was programmed to spend most of its time in a low-power, sleep mode. Periodically, the microcontroller would wake up and read the current time from the real-time clock. If it was time to take a measurement, the microcontroller would power the measurement circuit, otherwise it would go back to sleep. At one-hour intervals, the measurement circuit on the microSD shield was enabled, and the soil-moisture sensors were read and data stored to the microSD card.

To properly read a Watermark 200SS sensor, an alternating current source is recommended in order to avoid polarizing the sensor with a prolonged direct-current excitation, which can influence sensor measurements and degrade the sensor over time. The microcontroller can only supply a direct-current excitation, however, so a pseudo-alternating current source was created by rapidly switching the polarity of the direct-current voltage sent to power the sensor, and the sensors were then read under each polarity. Digital pin 6 was first set high (a voltage level of 5 V) and pin 5 was set low (a voltage level of 0 V) so that current flowed through the half bridge in one direction (seeFigure 3). The output voltage, Vout, between the 10 kohm resister, R, and the Watermark sensor, Rwm, was measured with an A/D converter, and the sensor resistance was calculated using the voltage-divider relationship, Vout = R/(Rwm + R)*5 V. The polarity of the half bridge was then switched by setting pin 5 high and pin 6 low, so that current flowed in the opposite direction, and output voltage was again measured and sensor resistance calculated. This was repeated five times, and an average resistance was calculated.

To arrive at the sensor's final output, namely the matric-potential of the soil, in kPa, a calibration equation is required to convert sensor resistance to matric potential. Much work has been done calibrating and verifying the Watermark 200SS sensor [14-16], and several calibration equations have been proposed. The equation of Shock et al. [16] was chosen, written as SWP = (4.093 + 3.213 Rwm)/(1 − 0.009733*Rwm − 0.01205*Tsoil), where SWP is the soil-water potential (kPa), Rwm is the sensor resistance (ohms), and Tsoil is the soil temperature (°C). While sensor performance has been shown to vary slightly with temperature, and a temperature-correction factor is included in the calibration equation, soil temperature was not measured and, instead, a constant temperature of 25 °C was used. To improve accuracy of sensor readings, a soil-temperature sensor could be added to the datalogger circuit and actual temperature measurements input to the calibration equation.

Following sensor measurements, power was sent to the microSD card circuit, and the data were stored to the microSD card. Data were stored as ASCII text, separated by spaces, in a plain text file, and consisted of six values; a datalogger board identification number, date (month/ day/year), time of day (hour), sensor #1 reading (kPa), sensor #2 reading (kPa), and sensor #3 reading (kPa). The microcontroller then turned all power off to the microSD shield and returned to low-power, sleep mode.

Data

Thirty soil-moisture sensor dataloggers were constructed and deployed to monitor soil-moisture status in experimental research plots at the USDA

Agricultural Research Service's Jamie Whitten Delta States Research Center at Stoneville, Mississippi USA. Research plots planted to soybean and cotton were instrumented with soil-moisture sensors and Arduino-based dataloggers. At each instrumented site, Watermark sensors were installed at three depths; 15-, 30-, and 60-cm, below the soil surface. The sensors were connected to a datalogger, and the datalogger was placed inside a weatherproof plastic enclosure attached to a wooden stake driven into the ground. The datalogger was turned on, and collected sensor data at one-hour intervals throughout the entire growing season. Periodically, each site was visited to download data from the microSD card to a portable tablet computer. The text data files were then returned to the office, uploaded to a desktop computer, and input to a spreadsheet for analysis and viewing. Typical data, from one site over a sevenweek period following planting in 2011, are shown in **Figure 4**.

Soil-water potential values near 0 indicate very moist soil conditions, with soil-water levels decreasing as the water-potential values become more negative. Hourly data from the three sensors were input to a spreadsheet, and the average of the three sensor readings was calculated. The average values were used to determine when an irrigation was needed. When the average values reached a threshold value of −50 kPa, an irrigation was scheduled. In **Figure 4**, soil-water levels decreased early in the season as the growing crop extracted water until rainfall occurred on 6/21, rewetting the soil. As soil water was used by the crop, the levels dropped until reaching −50 kPa, and two irrigations were required. Evident in the data are differences in water use with depth in the soil profile. Early in the season, changes in water potential were slower at 30 cm than at 15 cm, and much slower at 60 cm, suggesting more active roots in the shallower depths. As the season progressed, water-use rates increased at the 30-cm depth, and later at the 60-cm depth, suggesting increases in root activity and water extraction.

Ultrasonic Water-Level Datalogger

Fluid levels are measured in a variety of applications; fuel tanks, water reservoirs, and irrigation canals, for example. Evaporation pans are used to estimate the evaporative demands of the atmosphere in order to determine crop water use and soil evaporation rates for input in water-balance and evapotranspiration studies, and to assist in irrigation scheduling. A datalogger was de- veloped to automate the measurement of the depth of water in an evaporation pan using an ultrasonic distance sensor.

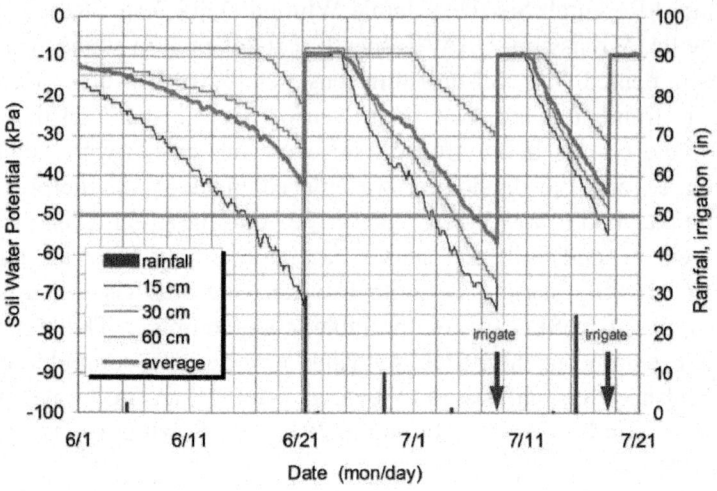

Figure 4: Data collected with the soil-moisture datalogger for a seven-week period in 2011.

Hardware

The ultrasonic water-level datalogger circuit is based on that of the soil-moisture datalogger, and incorporates many of the same circuit components. The same Ardinocompatible microcontroller board was used, and was modified in the same manner to supply a stable power source and enable battery-powered operation. The same microSD/prototyping shield and real-time clock components were also used. A schematic of the ultrasonic water level datalogger is shown in **Figure 5**.

An ultrasonic distance sensor, model SRF-04 (Devantech Ltd., Norfolk, UK), interfaces with the microcontroller via three digital pins; power, trigger, and echo pulse. The sensor consists of two ultrasonic transducers, one to send an ultrasonic pulse and one to receive the pulse's echo. To make a measurement, power is supplied to the sensor, and a measurement is initiated by sending a brief signal to the trigger pin, which causes an ultrasonic pulse to be sent. The microcontroller then begins monitoring the echo pulse pin, and measures the length of time it takes to receive an echo signal.

A temperature sensor was added to measure the air temperature of the environment. The LM35 analog temperature sensor outputs an analog-voltage signal in proportion to its temperature. The signal is input to one of the microcontroller's A/D converters, and a calibration equation supplied by the manufacturer is used to convert the voltage signal to temperature.

Software

The microcontroller program for the ultrasonic waterlevel datalogger used many of the same routines written for the soil-moisture datalogger. The microcontroller wakes periodically from a low-power sleep mode to read the real-time clock and determine if it is time to take measurements.

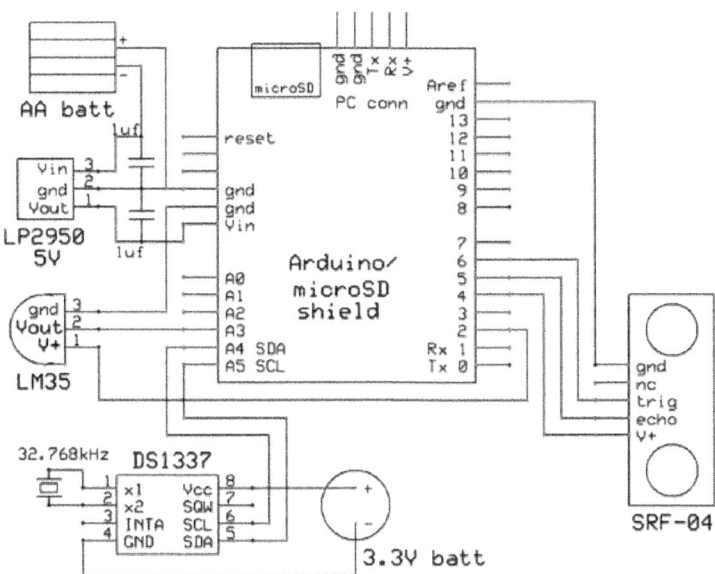

Figure 5: Schematic for ultrasonic water-level datalogger.

If so, measurements are taken and recorded, otherwise, the microcontroller goes back to sleep.

At each measurement interval, the microcontroller turns on the temperature sensor and makes an air temperature measurement. The ultrasonic sensor is then powered, a trigger signal is sent, and the time for an echo signal to return is measured. The time for the echo to return is then used to calculate the distance, based on the speed of sound, between the sensor and the surface upon which the ultrasonic pulse impacted. The speed of sound, however, is strongly dependent on the air temperature, and slightly affected by humidity [17], and can be corrected to improve the accuracy of distance measurements using the relationship $v = 331$ m/s $+ 0.6$ m/s/C$*$T, where v is the speed of sound (m/s) and T is the air temperature ($^{\circ}$C). To make a distance measurement, the air temperature measurement is first used to correct the speed of sound value. The speed of sound and the time taken to return the pulse echo are then used to calculate the distance from the sensor to the reflecting surface. This distance is then subtracted from the distance of the sensor to the bottom of the evaporation pan, measured previously when installing the ultrasonic sensor, to determine the depth of water in the pan.

Following air temperature and water level measurements, the data, microcontroller board identification number, and date and time are written to the microSD card. The microcontroller then turns off power to the circuit and returns to low-power, sleep mode.

Data

Two ultrasonic water-level dataloggers were constructed and installed in summer 2011 and operated for a threemonth period. The sensors were installed on an evaporation pan approximately 300 mm above the bottom of the pan. Sensor measurements were recorded at one-hour intervals, and the data were periodically downloaded from the microSD card during periodic site visits. During site visits, manual measurements of the water level were made by inserting a steel ruler into the evaporation pan and reading the depth of water. The depth of water in the pan varied between 70 and 195 mm, decreasing as water evaporated in response to the environmental demand and increasing due to rainfall and periodic manual refilling.

Data collected during a four-day period with one ultrasonic sensor datalogger are shown in **Figure 6**. Data include air temperature, raw depth (before correcting the speed of sound for temperature) and temperature-corrected depth, and manual measurements of the water levels. Large increases in apparent depth of water can be seen in the raw sensor readings each morning beginning around 6:00, as the sun rose and air temperature increased rapidly. The raw depths also continued to appear to decrease after sunset, when evaporation would be expected to cease. Correcting the speed of sound for air temperature mostly eliminates these errors, resulting in expected changes in water level, decreasing during daylight hours and minimal changes during nighttime. An increase in depth can be seen in response to a manual addition of water to the pan.

Accuracy of ultrasonic measurements was determined by comparing water levels measured with the ultrasonic sensors to those measured manually. Manual depth measurements were made 18 times, at varying times throughout the three-month period and at varying times of day. Manually measured water levels ranged from 75 to 158 mm. Comparison of measurements from the two ultrasonic sensors is shown in **Figure 7**, and indicates a very good agreement with the manual measurements, with a standard error of measurements of approximately 2 mm.

Environmental Datalogger

An Arduino-based datalogger can also be built using a custom printed circuit board (PCB) rather than starting with a commercially available board.

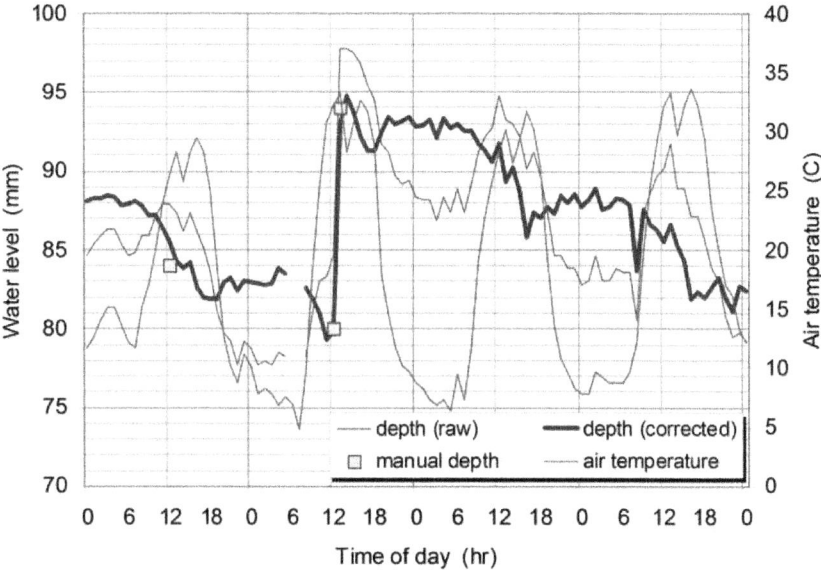

Figure 6: Hourly data collected with one ultrasonic water-level datalogger during a four-day period.

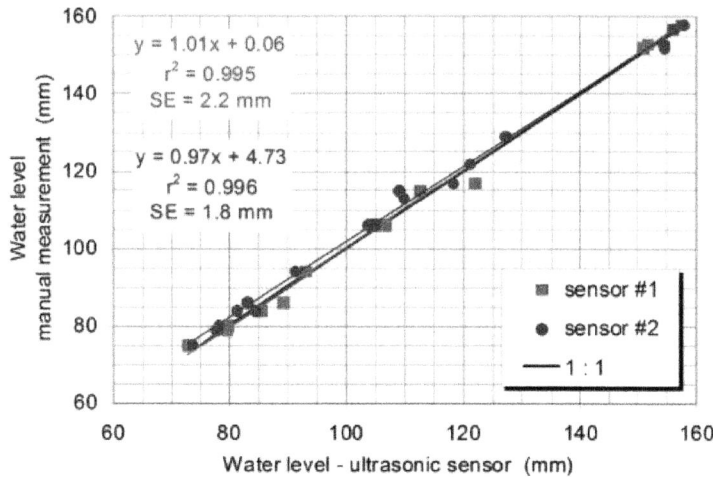

Figure 7: Comparison of manual versus automated measurements for two ultrasonic water-level dataloggers.

A datalogger was designed and fabricated to collect environmental data in a forested setting. The datalogger was de signed to accommodate a variety of sensor types, but was primarily intended to measure soil moisture and air temperature. Rather than developing a system around a commercially available Arduino board, a custom PCB was created which contained the Arduino microcontroller and other components.

Custom Circuit-Board Design

Creating a new PCB requires some additional skills but is a viable approach for many users. The main advantages of creating a custom PCB include the selection of specific components for the particular project, the creation of a board with a custom layout or size/configuration, and the reduction, in some cases, in total cost.

The minimum components needed to create an Arduino board include the ATmega328 microcontroller, a resonator (self-contained oscillator circuit), a reset button, a voltage regulator, connectors for a battery pack, computer interface, and a few resistors and capacitors. ATmega328 microcontroller chips are available preprogrammed with the Arduino-system's bootloader, enabling the use of the Arduino IDE to create and upload programs to the microcontroller. Additional components for most datalogger projects include a real-time clock/ calendar, a memory device, and one or more light-emitting diodes (LEDs) to indicate the operational status of the datalogger. All of these components are readily available as through-hole components which can be soldered to the PCB with a soldering iron.

The process of creating a custom circuit board begins with circuit and PCB design. Several software packages are available, some in freely available, open-source versions, to design the electrical schematic and then lay out the circuit on a PCB. A graphical user interface simplifies design, and the software creates a set of files in formats standardized for PCB manufacturing, which can then be transmitted to a PCB manufacturer. The manufacturer produces the bare PCB, and the final board is constructed by soldering the components to the PCB by hand.

Hardware

A board was designed using the freely available Design-Spark PCB software (www.designsp ark.com/pcb). The circuit was designed using a graphical schematic view, in which connections between circuit components are created but the actual size, shape, and layout of the components are unimportant. This schematic is then transferred to a printed circuit board layout, where the software suggests the physical layout and connecting traces of the components. The user is able to modify the layout as desired, to create a PCB that is easy to assemble, or which fits certain dimensional or other constraints.

The resulting board design, with dimensions of approximately 60 × 90 mm, was then electronically transmitted for fabrication using SeeedStudio's Fusion PCB service (www.seeedstudio.com/propa gate). Dataloggers were assembled by soldering circuit components to the custom PCB, with each datalogger requiring approximately 20 minutes to complete. A list, with approximate cost of components, excluding sensors, is provided in **Table 2**. The original design layout is shown in**Figure 8**(a), with resulting bare printed circuit board and finished datalogger board shown in Figures 8(b) and 8(c), respectively.

Dataloggers were deployed in the field along with air-temperature and soil-moisture sensors. Air temperature was measured using a DS18B20 12-bit digital temperature sensor (Maxim Integrated Products, Inc., Sunnyvale, CA USA). The sensor uses the 1-Wire communication protocol to transfer measurements to the microcontroller, and contains an internal 18-bit A/D converter which provides temperature measurements with a resolution of 0.06 C. Soil-moisture measurements were made using an EC-20 capacitive sensor. The microcontroller provided an excitation voltage to power the sensor via a digital IO pin, and measured the analog output voltage with a built-in A/D converter.

A battery pack consisting of 5 AA alkaline batteries enabled long-term remote operation by ensuring adequate voltage as the batteries discharged.

SUMMARY

Advances in electronic technologies, microcontrollers, and sensors offer researchers a variety of new and inex- pensive sensing, monitoring, and control capabilities.

Table 2: List of materials for environmental datalogger

Main components	Part number	Supplier	Cost $
Microcontroller with bootloader	ATmega328	Sparkfun Electronics	6
Printed circuit board		SeeedStudios	3
Memory chip	24LC512	Microchip	4
Real-time clock/calendar	DS1307	Maxim Integrated Products	1
Oscillator	32.768 kHz	Citizen America	1
Regulator	LP2950	National Semiconductor	1
Screw terminals			4
Miscellaneous (capacitors, resistors, connectors, LEDs)			6
		Datalogger Total	26

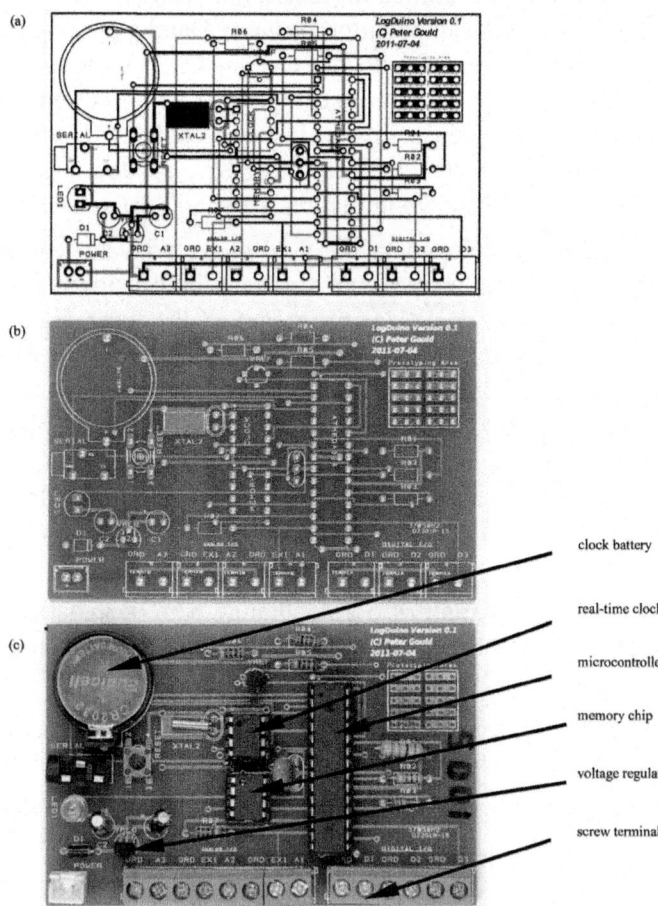

Figure 8: Arduino-based datalogger fabricated on a custom printed circuit board: top, (a) circuit-board layout; middle; (b) bare circuit board; bottom; (c) completed board with components installed.

The concept of open-source hardware, in which hardware designs, software programs, and development efforts are made freely available to all, help facilitate and expand the adoption of these capabilities. The open-source hardware Arduino development platform has great potential for implementation in scientific research applications, and can empower researchers with flexible, inexpensive tools for expanding their data-collection, automation, and control capabilities.

DISCLAIMER

Mention of a trade name, proprietary product, or specific equipment does not constitute a guarantee or warranty by the United States Department of Agriculture, and does not imply approval of the product to the exclusion of others that may be available.

REFERENCES

1.   F. H. Moody, J. B. Wilkerson, W. E. Hart and N. D. Sewell, "A Digital Event Recorder for Mapping Field Operations," Applied Engineering in Agriculture, Vol. 20, No. 1, 2004, pp. 119-128.

2.   K. A. Noordin, C. C. Onn and M. F. Ismail, "A Low-Cost Microcontroller-Based Weather Monitoring System," CMU Journal, Vol. 5, No. 1, 2006, pp. 33-39.

3.   D. K. Fisher, "Automated Collection of Soil-Moisture Data with a Low-Cost Microcontroller Circuit," Applied Engineering in Agriculture, Vol. 23, No. 4, 2007, pp. 493-500.

4.   G. Vellidis, M. Tucker, C. Perry, C. Kvien and C. Bednarz, "A Real-Time Wireless Smart Sensor Array for Scheduling Irrigation," Computers and Electronics in Agriculture, Vol. 61, No. 1, 2008, pp. 44-50. doi:10.1016/j.compag.2007.05.009

5.   D. K. Fisher and H. Kebede, "A Low-Cost Microcontroller-Based System to Monitor Crop Temperature and Water Status," Computers and Electronics in Agriculture, Vol. 74, No. 1, 2010, pp. 168-173. doi:10.1016/j.compag.2010.07.006

6.   Arduino, "An Open-Source Electronics Prototyping Platform," 2012. http://www.arduino.cc

7.   D. Bri, H. Coll, M. Garcia and J. Lloret, "A Multisensor Proposal for Wireless Sensor Networks," 2nd International Conference on Sensor Technologies and Applications, Cap Esterel, 25-31 August 2008, pp. 270-275.

8.   L. Buechley and M. Eisenberg, "The LilyPad Arduino: Toward Wearable Engineering for Everyone," Pervasive Computing, Vol. 7, No. 2, 2008, pp. 12-15.doi:10.1109/MPRV.2008.38

9.   J. Zhang, S. K. Ong and A. Y. C. Nee, "Design and Development of a Navigation Assistance System for Visually Impaired Individuals," Proceedings of the 3rd International Convention on Rehabilitation Engineering & Assistive Technology, Singapore, 22-26 April 2009.

10.   N. W. Bergmann, M. Wallace and E. Calia, "Low Cost Prototyping System for Sensor Networks," 6th International Conference on Intelligent Sensors, Sensor Networks and Information Processing, Brisbane, 7-10 December 2010, pp. 19-24.doi:10.1109/ISSNIP.2010.5706802

11.   D. Gordon, M. Beigl and M. A. Neumann, "Dinam: A Wireless Sensor Network Concept and Platform for Rapid Development," 7th International Conference on Networked Sensing Systems (INSS), Kassel, 15-18 June 2010, pp. 57-60. doi:10.1109/INSS.2010.5573290

12.   J. Sarik and I. Kymissis, "Lab Kits Using the Arduino Prototyping Platform," Frontiers in Education Conference, Washington DC, 27-30 October 2010, pp. 1-5.

13.   M. Thomas, "In situ Measurement of Moisture in Soil and Similar Substances by 'Fringe' Capacitance," Journal of Scientific Instrumentation, Vol. 43, No. 1, 1966, pp. 21-27. doi:10.1088/0950-7671/43/1/306

14.   S. J. Thomson and C. F. Armstrong, "Calibration of the Watermark Model 200 Soil Moisture Sensor," Applied Engineering in Agriculture, Vol. 3, No. 2, 1987, pp. 186- 189.

15.   E. P. Eldredge, C. C. Shock and T. D. Stieber, "Calibration of Granular Matrix Sensors for Irrigation Management," Agronomy Journal, Vol. 85, No. 6, 1993, pp. 1228-1232.doi:10.2134/agronj1993.000219620085000 60025x

16.   C. C. Shock, J. M. Barnum and M. Seddigh, "Calibration of Watermark Soil Moisture Sensors for Irrigation Management," Proceedings of the International Irrigation Show, San Diego, 1-3 November 1998, pp. 139-146.

17.   D. A. Bohn, "Environmental Effects on the Speed of Sound," Journal of the Audio Engineering Society, Vol. 36, No. 4, 1988, pp. 223-231.

Chapter 3

FAULT HANDLING IN PLC-BASED INDUSTRY 4.0 AUTOMATED PRODUCTION SYSTEMS AS A BASIS FOR RESTART AND SELF-CONFIGURATION AND ITS EVALUATION

Birgit Vogel-Heuser, Susanne Rösch, Juliane Fischer, Thomas Simon, Sebastian Ulewicz, Jens Folmer

Institute of Automation and Information Systems, Technische Universität München, Munich, Germany

ABSTRACT

Industry 4.0 and Cyber Physical Production Systems (CPPS) are often discussed and partially already sold. One important feature of CPPS is fault tolerance and as a consequence self-configura- tion and restart to increase Overall Equipment Effectiveness. To understand this challenge at first the state of the art of fault handling in industrial automated production systems (aPS) is discussed as a result of a case study analysis in eight companies developing aPS. In the next step, metrics to evaluate the concept of self-configuration and restart for aPS focusing on real-time capabilities, fault coverage and effort to increase fault coverage are proposed. Finally, two different lab size case studies prove the applicability of the concepts of self-configuration, restart and the proposed metrics.

INTRODUCTION

In the course of Industry 4.0, intelligent products and production units are implemented. This includes production units with inherent capabilities which adapt (also structurally) flexibly in response to changing product requirements [1] or in case of failures of a partial component in order to stay or become operable again and increase the Overall Equipment Effectiveness (OEE) [1] . How can the adaptability, i.e. reconfiguration and restart of these automated production systems be evaluated and are these strategies already implemented in industry or is there still a huge gap to be bridged?

In the reference architecture (RAMI [2]), attributes and requirements to Industry 4.0 components are specified and allow the evaluation of different solutions offered under the heading of Industry 4.0, but also serve for further development of attributes and metrics. In the following, one aspect of these requirements, namely adaptivity and selected metrics for adaptivity, particularly in the event of a fault, are presented to allow a comparison of different adaptivity concepts. Despite adaptivity already being discussed in academia, industrial automated production systems (aPS) lack such a concept. In systems design, consistency of models from different disciplines for specific purposes (fault analysis, safety) is required, but needs further analysis and support in the future [3] . To elaborate the challenge to realize such concepts in real world industrial applications, the state of the art in fault handling and software architecture as a basis for adaptivity after a fault is given, derived from eight industrial case studies (cp. Section 3).

The remainder of this paper is structured as follows. At first, the state of the art in engineering and operation of Programmable Logic Controller (PLC)-based aPS is given highlighting challenges and weaknesses in engineering processes, platforms and languages, domain specific extra functional challenges, concepts for recovery in case of faults as well as metrics to measure fault-recovery as a prerequisite for adaptivity. Next, an analysis based on eight case studies has been conducted to capture real world software architecture including concepts for fault-handling. Fault handling is chosen as it is strongly related to modes of operation in aPS and differs significantly from classical software also in embedded systems.

In aPS, operator personnel often has to fix faulty situations manually because the possible faults and fault combinations are manifold [3] and often need manual mechanical intervention by an operator.

The case studies reveal a software architecture with five levels of hierarchy with mostly hierarchical fault handling mechanisms on the one hand, but a lack of any automatic fault recovery as a prerequisite for adaptivity on the other hand. Instead, fault handling requires human intervention, therefore faults detected by the PLC need to be addressed in the human machine interface (HMI). We identified three different interface concepts in the case studies between PLC fault handling and HMI.

Despite the lack in realization in industrial aPS up to now, adaptivity concepts have been developed in academia and implemented in lab-size demonstrators for Industry 4.0, e.g. myJoghurt. To evaluate these concepts comparatively and measure the needed additional modelling or programming effort and the required real time behavior, selected metrics are introduced (in Section 4). Four different adaptivity concepts are applied and evaluated to lab size demonstrators in Section 5. The evaluation section closes with a comparison of the metrics' results interpreting the significance of the metrics values in comparison. In Section 6, the summary and outlook for future development are given.

STATE OF THE ART—CHALLENGES AND WEAKNESSES

This section starts with an overview on the state of the art regarding model-driven engineering, platforms and programming languages for aPS, introducing modes of operation and finally alarm handling.

Development of Runtime Environments and Their Domain Specific Challenges of Programming Languages for aPS

After a short introduction to the specific characteristics of Programmable Logic Controllers, which are the standard industrial platform, and IEC 61131-3, which is the current programming standard, actual architectural approaches are introduced. Subsequently, the various challenges of developing software for aPS and challenges in the domain of aPS are described for a better understanding of maintenance issues of PLC code.

A. Platform, programming languages and software architecture in the aPS domain

Programmable Logic Controllers (PLC) are characterized by their cyclic data processing behavior, which can be divided into four steps. At the beginning of a cycle, the PLC reads the input values of the technical process, which are provided by sensors, and stores them in a process image. Subsequently, the PLC program is executed with the stored values and afterwards the output values are written, which control the actuators that influence the technical process. At last, the PLC waits until the cyclic time has elapsed. In a worst case scenario, if a fault occurs right after the input values have been read, the reaction time of the PLC is two times the cycle time.

The IEC 61131-3 programming standard for PLCs consists of two textual languages—Structured Text (ST) and Instruction List (IL)—and three graphical languages—Ladder Diagram (LD), Function Block Diagram (FBD), and Sequential Function Chart (SFC). Furthermore, the standard defines three types of program organization units (POU) to structure PLC code and to enable reuse: programs (PRGs), function blocks (FBs) and functions (FCs). The main differences between these POUs are that in contrast to FCs, PRGs and FBs possess internal memory and that FBs can be instantiated. Tasks are used to define entry points (PRGs) into a plant's code, which are invoked depending on the defined cycle time of the task. The entry points (PRGs) then call other POUs which can execute code and sub-calls of further POUs. A regular PLC execution cycle consists of reading all inputs (sensors), triggering tasks depending on their cycle time and therefore their associated code (POU calls and subcalls), and finally writing all output variables (actuators). These cycles adhere to real time requirements, meaning that the defined cycle times of the tasks may never be exceeded.

Software engineering for aPS is still struggling with modularity, guidelines for appropriate software component sizes [4] - [6] and good practices for interfaces between these components. Katzke et al. [7] and Jazdi et al. [4] found different component sizes in aPS software (called granularity) and described the challenge to choose the best size and interface in between components for reuse and evolution. Cross component functions such as fault handling and modes of operation (manual, automatic) make the implementation of many modularity concepts difficult, which will be discussed in more detail in Section 3. Because in the plant manufacturing industry software engineering has been mostly project driven for decades, the challenge is to restructure legacy code from different projects with similar or even equal functionality. To make things worse, the different platforms (cp.Table 1) require software variants for the same functionality due to different IEC 61131-3 dialects.

Based on Katzke [7] the authors proposed a five level architectural model in [8] (cp. Figure 1). "A plant module resembles a whole production plant and, consequently, exists mostly in the plant manufacturing industry, but not in machine manufacturing industry. A plant module usually contains several facility modules, which re- present machines or plant parts such as a press or a storage system. Each facility module in turn consists of one or more application modules, which are machine parts that might be reused in other machines such as the material feed of a machine or the filling unit of a machine. Application modules are composed of basic modules which represent for example individual drives or sensors. Atomic basic modules represent the most fine-grained architectural level and refer to basic modules that cannot be decomposed into further module. The architectural levels can be used recursively, i.e., each level can consist of all module types of the more fine-grained levels." [8] . Vyatkin proposes a software architecture for distributed automation systems based on IEC 61499 [9] [10] . The resulting software shows a composite structure and consists of event-driven FBs, which are used to describe processes. Although first industrial applications confirm the standard's benefits, e.g., reduced time and effort to develop automation software, a high degree of code modularity and a high potential for reuse, the standard is not commonly used within industry at present.

Current research in the field of model-driven engineering (MDE) is mainly focused on developing new methods to support the development process of new software using modeling languages such as UML or SysML. Unfortunately, there is a big gap between existing legacy code on field level, i.e. PLC code, and the vision and attempt to introduce an MDE approach in industrial companies supporting a systematic maintenance and software evolution.

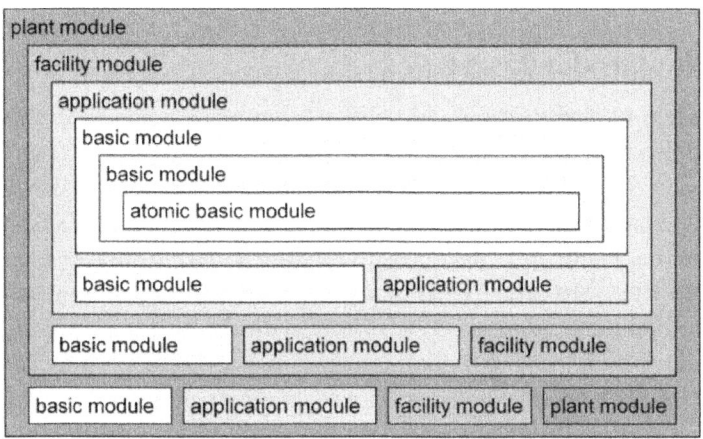

Figure 1: Architectural layers from [8] .

Table 1: Table classified case studies based on and enlarged from [8]

Description	Case study A	Case study B	Case study C	Case study D (1)	Case study D (2)	Case study E	Case study F (1)	Case study F (2)	Case study G	Case study H
Application domain	MT	MT	MT	MT		MT	PT	MT	PT/MT	MT
Area	M (automotive)	M (packaging)	P (consumer goods)	P (wood working)		M (packaging)	P (filling)	P (logistics)	P	M
PLC type and supplier	PLCopen compliant	PLCopen compliant	Siemens S7	Siemens S7, Rockwell		PLC open compliant	Siemens 1500, Rockwell		Siemens S7, Rockwell	PLCopen compliant
Languages	ST, SFC	FBD, ST, SFC	LD	LD, IL		ST	IL	FBD, IL, SCL	IL, FBD	SFC, ST (OO)
Number of components	127	59 + 140	168	133	278	330	356	296	na	na
Number of supplier's library components	18	140	(16)	-	-	274	74	98	na	na
State machine per module	-	OMAC	-	-	-	OMAC	-		OMAC	Part of engineering environment
Restart after fault	-	(O)	-	-	-	(O)	-	-	na	na
Size of memory [MB]	1.9	14	12.3	0.150	0.580	4.2	0.423	0.218	na	na
Fault handling	I, III	I	II	III	III	I	II		na	na
Maintenance staff	Eng, Tec	Eng, Tec	Tec, SW	Eng, Tec, SW		Tec	Eng, Tec		Eng, Tec	Eng, Tec

PT—process technology, MT—manufacturing technology, P—plant, PU—plant unit, M—machine, Eng—Engineers, Tec—Technicians, S—Skilled Workers, na—not available, (O)—partially realized.

To bridge this gap, at first code refactoring and building of appropriate software components is required. Bonfè et al. introduce the concept of mechatronic objects to enhance modularity of the software, which can be represented in control programs by FBs. While the structure of the aPS is modeled using UML class diagrams, the behavior can be defined with UML state diagrams [11] . Although various models for software architecture have been developed, accepted software architectures are missing in aPS up to now.

B. Constraints due to size

While machine manufacturers are able to commission and test their products before delivering it to the customer, plant manufacturers are unable to commission the entire plant in their own facilities due to the weight and the size of the components. A press for example is too large to be transported fully assembled on a ship and thus needs to be transported in smaller pieces and be assembled afterwards at the customer's site. Consequently, commissioning and startup of complete plants is done on site with the customer pressing to start production, which leads to high time pressure. Furthermore, it is common that the start-up staff on site includes only technicians or skilled workers instead of the application engineers who developed the PLC software for the plant. Therefore, the communication between development engineers in the office and technicians on site needs to be coordinated which can be challenging in a company that distributes its products globally due to different time zones and time pressure with high penalties.

C. Extra functionality challenges: modes of operation and fault handling

In addition to implementing the control functions that are carried out by an aPS, also aspects like different operation modes or visualization must be taken into account.

According to Güttel et al. [12] the main operations of function blocks include:

- Automatic mode: defines the behavior of a machine part in automatic mode.
- Setup mode: defines the behavior of a machine part in setup mode. In this mode, the drive of a machine part moves as long as the manual input is active. In this mode, no interlocks are active.
- Manual mode: defines the behavior in manual mode. In this mode, the interlocks are active.

- Semi-automatic mode: defines the behavior of a machine part in semi-automatic mode.
- Initialize: defines the behavior during the initialization of the machine part.
- Shut down: defines the behavior during shutdown.
- Save stop: defines the actions which are necessary to reach a safe state.

The different operation modes need to be implemented, which may be realized as additional automata or different branches with the other automata according to Fantuzzi et al. [13] . In food & beverage a more sophisticated standard is used, i.e. the OMAC standard or in Germany the Weihenstephan Standard [14] .

The OMAC State Machine [15] (cp. Figure 2) is part of the widespread PackML standard, which pursues the objective to bring operational consistency to a packaging line, especially if it consists of packaging machines from different vendors. The PackML standard defines the OMAC State machine with 17 states consisting of acting and waiting states in which acting states represent activities like starting and waiting states identify the reaching of a set of conditions e.g. Held. The OMAC State Machine is responsible for identifying valid state transitions depending on the actual OMAC state and specified state transition conditions. If a state change occurs, a suitable function is called, that is implemented by the machine vendor or integrator.

Fault Recovery and Restartability in aPS

In the following four different concepts for fault recovery to increase OEE by adaptation are introduced in principal. Fault recovery after a device fault may be realized automatically in case of redundant information/devices available. The redundant information may originate from an additional device and a Fault Coverage Analysis (cp. Section 5.1 according to [16]) providing the faulty information or from a redundancy model designed pre- runtime and calculated during runtime using other process information (cp. Section 5.3, [17]). Both scenarios will be discussed and evaluated with the metrics to be introduced during the evaluation examples. The third scenario focuses on restart after emergency shutdown remembering the position of the work piece to restart properly, which is modelled accordingly during the design phase (cp. Section 5.3). The fourth scenario examines restartability after machine stop [18] to ease the operator's task and reduce operator interaction. This approach is based on an OMAC State Machine (cp. Section 5.4).

Concerning the elimination of faults our work is based on prior results by Schütz et al. [17] about dynamic reconfiguration and, if present, an indirect replacement sensor being used to substitute a defect sensor or actua- tor with a virtual sensor, but with less accuracy. Schütz et al. [19] have developed a model-based approach for this based on SysML and are able to automatically generate executable PLC-code for the knowledge base from this model.

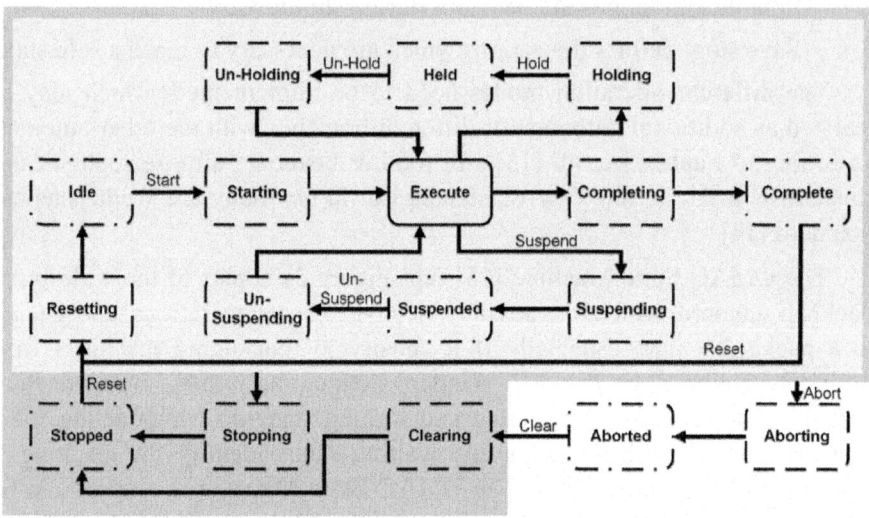

Figure 2: OMAC state machine including waiting states (dashed-dotted framing), acting states (dashed framing) and the dual state execute, which is both a waiting and an acting state (bold framing).

Priego et al. developed an architecture that allows the definition, generation and update of a PLC's software architecture during runtime in response to modeled changes on the plant layout, the product to be manufactured or the control hardware. [20] . Andersson et al. [21] [22] address the challenge to automatically derive operation sequences for restarting manufacturing systems (and here especially assembly cells typically used in the automotive industry). In context of these works, restarting is the procedure to "resynchronize the control system and the physical system, such that the production can be restarted and eventually complete" [23] . These concepts are required because the intended nominal production is not performed due to an error during operation. The concept is based upon the concept of self-contained operations which contain information about their potential sequential relations to other operations and, accordingly, can be arranged to so-called sequences of operations [24] . For modeling possible operation sequences, Andersson et al. rely on extended, deterministic, finite automata and supervisory control theory formalisms (i.e. controller synthesis) for deriving a desired restart strategy. Nevertheless, flexible manufacturing systems facilitate the production according to multiple operation sequences [25] . Therefore, Bengtsson et al. [26] [27] extended the modeling concept to multiple operation sequences. Here, identification of an earlier state to restart the system is more complex and requires efficient algorithms. Based on this approach, Bergagård et al. [18] [25] generalize the approach for restartability of Andersson et al. by extending the concept to multiple resources to be restarted (i.e. multiple operation sequences). This broadens on the one hand the set of manufacturing systems which can be addressed by this approach and on the other hand, by means of the generalization, also more efficient synthesis algorithms can be applied [23] .

Metrics for Adaptive aPS

As characteristics for Quality of Service real-time capability, reliability and the possession of the required QoS- attributes are named in RAMI. If solely the Overall Equipment Effectiveness (OEE) is used as metric for reconfiguration during runtime in case of failure, statements on the continuity of production despite a fault and the following possible output and its quality can be made. In the following already existing metrics for flexibility and adaptivity are summarized as a basis for the derivation and presentation of the developed metrics presented in this paper. Ladiges et al. [28] present metrics for flexibility divided into machine flexibility, process flexibility, routing flexibility and operation flexibility for discrete production processes. Thereby, the absolute machine flexibility is defined as number of operations that can be performed without any form of manual intervention. In contrast to that, Ladiges et al. [28] calculate the relative machine flexibility as a machine's number of operations without manual intervention, which can be performed in reference to all tasks within the facility, for example, tasks which can be taken on by other machines. Gronau et al. [29] establish an evaluation scheme for adaptivity within the scope of business-specific versatility and distinguish between five levels: no adaptation possible, adaptable via add-ons, adaptable via modification, adaptable via parameterization and adaptability via self-con- figuration. In the following we focus on adaptability via self-configuration in case of faults. Wiendahl et al. [30] define characteristic features for reconfigurable production and assembly systems e.g. modularity and scalability. Raibulet et al. [31] present several metrics for adaptivity from the perspective of computer science, classified according to architecture metrics, structural metrics, performance metrics and interaction metrics. Thereby, the main focus of architectural metrics lies on the costs of adaptivity. Structural metrics emphasize code changes due to the adaptivity and thus the additional required code entities while performance metrics rate the response delay caused by the adaptations as well as the response quality. The latter aspect is, in the following, adapted for production automation, formulated and evaluated. Regarding interaction Raibulet et al. [31] distinguish between users, which are operators in our case, maintenance personnel and administrative interaction (engineering).

RESULTS OF THE CASE STUDY ANALYSIS ON FAULT HANDLING AND SOFTWARE ARCHITECTURES WITHIN INDUSTRIAL AUTOMATION SOFTWARE FOR APS

In industrial practice in machine and plant engineering, initially the general software architecture and especially handling of faults was analyzed in order to examine how reconfiguration and recovery in case of faults is already or will be implemented in future and which challenges thereby arise and finally how to evaluate different strategies with metrics.

In cooperation with eight different companies in the field of machine- and plant engineering, ranging from special purpose machine manufacturers (Case studies A, B, E and H) to plant manufacturing industry (Case studies C, D, F and G) (cp. Table 1 for an overview of the case studies), a five-staged architecture model has been confirmed (cp. Figure 1) and, exemplary, the fault handling mechanisms were assigned to these five levels in order to analyze differences and similarities regarding fault handling and its integration into the software architecture (cp. Figure 3). Hereafter the characteristics are shortly introduced.

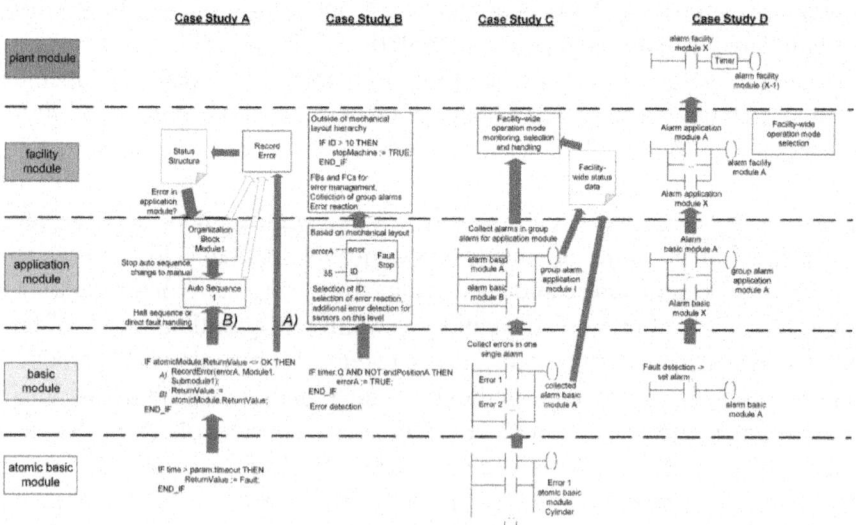

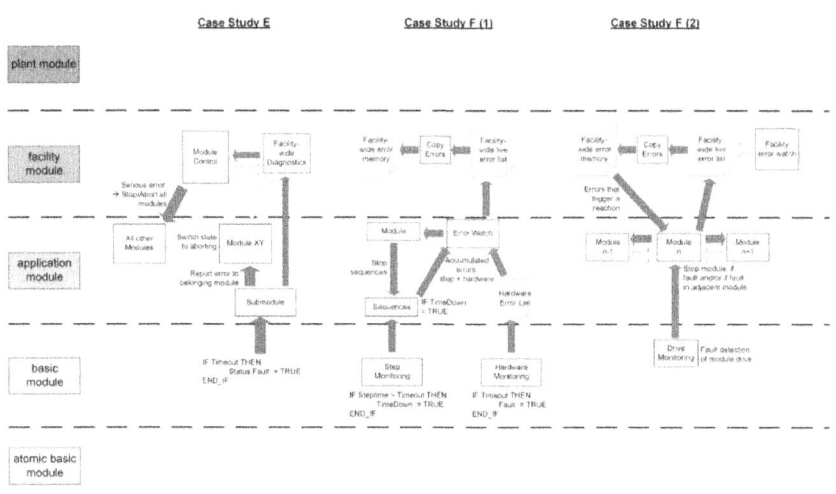

Figure 3: Fault handling in the case studies sorted into the five levels of the control software architecture (enlarged from [8]).

The case studies differ regarding the constraints due to size (row number of components, number of supplier's library components, and size of memory), i.e. delivering machine units or entire plants. Especially in plant manufacturing different platforms, i.e. PLC types need to be supported due to worldwide customers' requirements and as Vogel-Heuser et al. [3] already discussed on-site changes even by less qualified customer staff are mandatory to reduce downtime (row maintenance staff), which mostly leads to the usage of very simple programming languages, i.e. IL and LD. In some industry sectors, e.g. food & beverage domain specific standards occur to support the interlinking of different machines to one plant, e.g. OMAC and Weihenstephan Standard delivering a joined understanding of machine states. Regarding fault recovery and restart there is still a lack in industrial software, which will be discussed in the following.

The eight investigated case studies all implement hierarchical fault handling strategies with special forms. Considering real-time metrics is simple: The cycle time for fault detection amounts to one cycle. Isolating the fault may take several cycles if, for example, the group error has to be analyzed first. Rather than making the expense of fault isolation, which is occasionally fairly high (especially in case of alarm showers), the operator is provided with all alarms and following alarms connected to the fault. However, in order to carry out suggested reconfiguration and self-healing measures fault isolation is mandatory. Software-based, virtual sensors are so far not implemented in any of the industrial case studies, so that reconfiguration of functions via soft sensors, i.e. self-configuration according to Wiendahl [30] is not possible. In the following for four of the eight case studies fault handling is discussed in more detail to provide a deeper understanding of the domain.

Case study A—Special Purpose Machinery in Factory Automation for Automotive Sub-Component.

The first case study was conducted in the area of factory automation for production of automotive sub-com- ponents. In this domain, machinery is designed to customer's order (special purpose machinery) and thus the company has created a software structure which standardizes the software hierarchy within each unique machine. The structure is oriented at the hardware modules which are as follows: A production line (plant module) consists of multiple facility modules (each controlled by one PLC), which in turn consist of multiple application modules (larger sub-modules within the facility module, often representing one production process step). The structure definition allows for further subdivision of application modules into smaller application modules if reasonable. The most basic modules represent standard and special components such as pneumatic cylinders or vision systems.

The program of a facility module is usually limited to one task, which invokes the program block representing the program entry point. General program parts relating to the facility module and program parts relating to individual application modules of the facility are invoked from this entry point. The general program parts repre- sent functions such as diagnosis or initialization functions for the hardware controlled by this particular PLC. The entry point for each application module is a managing program block (program block related to decisions about sub-calls rather than direct relation to functionality of the application module), which is used to trigger sequences depending on the operation mode and initialization status. Invocation of further hierarchy layers using the same pattern can also be realized in the managing program block. Both general program parts and application module related structures invoke library functions, e.g., for communication to standard or special components.

Fault and alarm handling and subsequent error handling is initiated by atomic, basic or application modules. Atomic and basic modules include diagnosis functions for the connected modules, allowing simple fault detection, such as for stuck cylinders. Application module fault handling deals with errors in the process sequence which can result from failures of basic modules or the process itself. In addition, the application modules' managing program block monitors safety functions, such as safety door switches. Whenever a fault is detected, an alarm is collected in a centralized array alongside the facility module layer (cp. Figure 3, path A) and displayed on the HMI (cp. Figure 5, Fault Handling I). In case this fault is detected by a basic module, information directly relating to the component ID is included in the error message (cp. Figure 5, Fault Handling III). At the same time, the respective module changes into an error state and passes on the information to the parent module (cp. Figure 3, path B). This module in turn can implement failure handling algorithms, which are triggered according to the alarm. At the same time, the application modules' managing program block monitors status variables collected in the status structure and can halt automatic sequences in case of serious errors. In contrast to the concepts of some of the other case studies, the hierarchical module structure is partially bypassed by the centralized fault recording and status structure.

The machine state is realized through the operation modes and is loosely comparable with the OMAC states "execute", "aborting", "aborted", "resetting" and "starting". In automatic mode, the module is executed and aborted in case of errors resulting in a failure state ("aborted") which requires human interaction. The resetting is done by switching to manual mode, used to resolve process errors, followed by an automatic recalibration of the machine ("starting"), which is a prerequisite for returning back to automatic mode. Besides the facility module, each application module possesses these different operation modes, allowing for sub-states in the machine. While the OMAC state machine is not used, this procedure of switching between machine operation modes and thus states is standardized within this company.

Case study B—Machinery for Packaging Industry

The example in case study B consists of three tasks which invoke three PRGs. The task with the second highest priority and its corresponding program realize and call the main functions controlling the technical process and are strongly related to the mechanical layout of the machine. The mechanical layout consists of one facility module and six main application modules, which is directly reflected within the PLC control software architecture (with the exception of several additional software modules being used for facility-wide functions, e.g., for error management). Each application module includes similar subroutines, such as parameterization or axis movement, which are implemented differently and mostly include additional application and basic modules. Furthermore, encapsulated FBs and functions are used which stem from a supplier delivering and supporting axes in particular. The application conforms mostly to the ISA-S88 and implements the OMAC state machine model.

Fault and alarm handling may be considered as being handled in a hierarchical manner (cp. Figure 3). On the basic module level the main part of fault/error detection of hardware and faults stemming from the technical process is happening. This is only logical as, e.g., a pneumatic cylinder not reaching its end position, should be identified by the basic module pneumatic cylinder. It allows for easier reuse, as the mechanical representation has a direct software implementation. However, the error ID is assigned on the next higher level along with the decision on how the identified error should be handled (related to the severity of the error in rising order: only a warning is issued, the machine is immediately shut down). The advantage in this kind of setup is that errors are assigned to the correct application module (e.g., which pneumatic cylinder in which module is erroneous), making error identification easier. Furthermore, if more than one error occurs within one application module, but from different basic or sub-application modules, an analysis can be done and group errors, hinting more specifically at the cause of an error, may be identified. Group errors lead, depending on their severity, to the shutdown of the entire machine group as they often hint at specific problems in the specified group (area of a machine). In the next step the errors are analyzed by separate functions, apart from the application modules when considering the mechanical layout. The functions collect all errors and implement the error reaction. If several errors arrive, which individually would only be reported, the error reaction of shutting down may for example be set. This kind of setup allows for easy maintenance, as the overall error management functions have been standardized as libraries and can be reused within every machine. The error detection and error reactions are carried out mainly within the second task. The task with the lower priority executes functions related to the HMI (sending the alarms).

After an error that leads to a shutdown has occurred a function for restarting is available. If the calibration has not been impaired by the shutdown and the material is not entangled somewhere in the machine, the operator can decide to acknowledge the error and restart using the implemented function.

Case study D—Plant Manufacturer: logistics process

The storage system analyzed in case study D consists of five application modules (for a detailed description cp. [32]) and alarm handling within the storage is divided according to these five modules. Within the application module "general functions" a few general storage alarms, i.e., safety door alarms, are generated and also alarms of other application modules, such as "storage car" or "interface functions", are collected. If more than one storage car is used in the storage system, the application module "storage car" is reused and the alarms of the second car are also used for interlocking conditions in the first car's application module and vice versa. Depending on the severity of a detected fault (for example failure in application module "storage car 1"), the whole facility module "storage" or only parts of the storage system (storage car 1 stopped, but storage car 2 still operating normally) is switched to emergency mode. The shutdown of single application modules (e.g., "storage car 1") or the entire facility module "storage" according to the severity of the detected error is similar to the group errors described in case study B. If the facility module storage system is switched to emergency mode, the alarm is, additionally, collected on plant module level. After a delay time, the preceding facility modules, which transport production goods to the storage system, need to be stopped as well, due to a limited buffer for accumulated production goods. In summary, alarm handling in case study D is implemented in a hierarchical manner (cp. Figure 3).

Case study E–(Special) machinery in packaging

In this case study, a well-defined software architecture is used that corresponds mainly to the physical layout of the machine. The latter is divided in several modules on application level which are all initialized and controlled by a facility wide module control (cp. Figure 4). Each module has again one or more submodules and each submodule uses one or more basic elements. Furthermore, every module has got its own and independent state machine according to the OMAC standard [15] and passes on its state to its belonging submodules which—just like the modules—execute routines according to the given state. In these routines corresponding commands are given to the belonging basic elements. Additionally, the supervisory module control has got an own state machines well and is able to cause state switches of the modules by a defined set of commands via the respective interfaces.

Such a state switch can be induced by an occurring error, e.g. in a submodule, which is reported to a facility wide diagnostic module. The latter evaluates all reported errors and as a result sets an according diagnostic mode. Additionally, the module in which the error occurred immediately switches to the state aborting. Depending on the severity of the error, all other modules may have to be either stopped or aborted which is taken into account by the diagnostic module. Depending on the diagnostic mode, the module control subsequently sets a respective command to force all other modules to switch to the state stopping or aborting respectively. Hence, all modules have independent state machines but they are coordinated by the commands of the facility wide module control.

There is no full restart function available after the machine has been shutdown due to an error. However, axes that have been interrupted in their movement automatically move into a respective position from where a restart is possible.

After a more detailed analysis of the four case studies regarding fault handling discussing the concepts of restart and self-configuration in all cases, we can summarize that due to the character of the technical process and the resulting plants, implementing restart functions is not always possible.

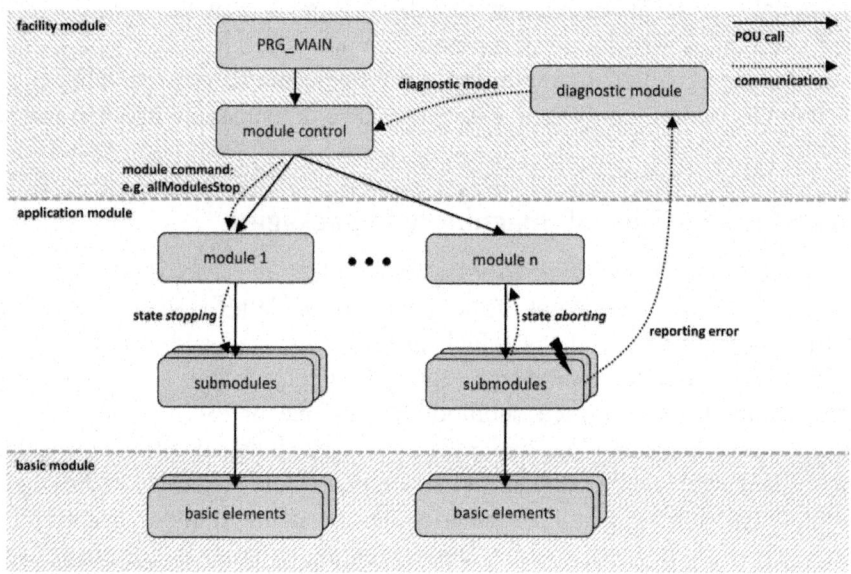

Figure 4: Software architecture and fault handling—case study E (figure caption).

Since some plants (such as case study D, F, G) contain continuous as well as discrete processes, a high number of interventions and different states arise and especially the continuous processes are highly interconnected. Furthermore, material properties often change during a continuous process and it is therefore not possible to return a plant into a previous recovery state with a defined material property, since once changed, an earlier material property cannot be retained.

Because of the missing restart and self-configuration abilities in industrial case studies, operators need to readjust the machines and plants into an appropriate state by manual intervention and via HMI interfaces. As interface to the operator three main variants to handle fault detection and fault handling with HMI (Human-Machine-Interaction) were identified (cp. Figure 5).

In variant I, faults are detected at the lowest level and collected in an array within a Function Block (FB Fault). Every fault that might occur in the machine has a unique, machine-wide fault ID. Furthermore, the entire machine is hierarchically divided into modules as depicted in Figure 5 with each of the modules and submodules having a unique identification and a unique instance number. After a fault is detected on submodule or module level, fault ID, module and submodule ID as well as module and submodule instance number are combined to form an unambiguous fault number, which is transferred to the Function Block FB Fault. By analyzing the final fault number, the fault type as well as its exact location within the machine can be determined. This approach to fault handling has, due to the unambiguous fault numbers, a high degree of flexibility (concerning, for example, the addition of further machine modules).

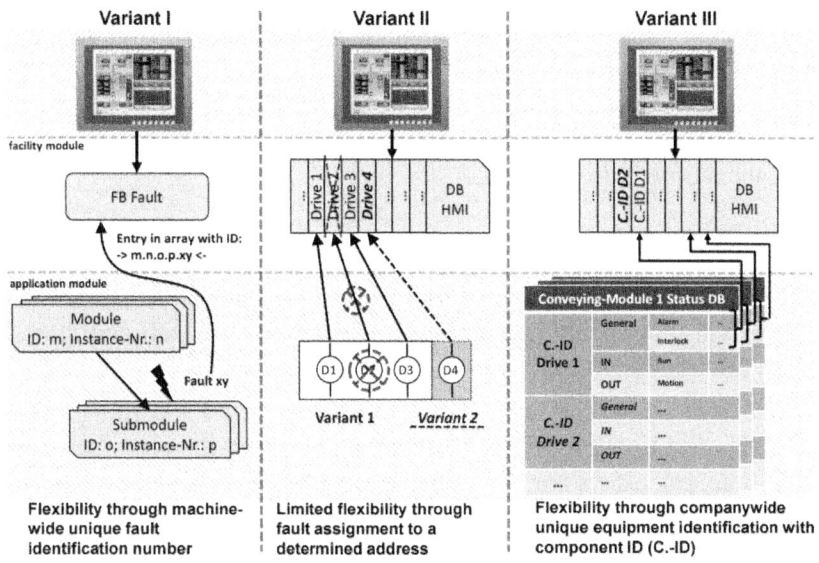

Figure 5: Identified variants of fault handling in between PLC and HMI interface as analyzed in the case studies.

In contrast to variant I, the fault ID in variant II is not based on the occurring fault and the according fault location, but each component of the plant, such as an actuator (e.g. drive) has a determined address within a data block DB HMI to transfer the component's alarms to. However, if the plant's components need to be altered or changed, addresses cannot be changed. This leads to unused memory when removing a component (cp. Variant II in Figure 5: removal of

Drive 2 leads to unused memory in DB HMI) and, moreover, the addition of a further component is only possible, if a spare address in the data block has been provided. Therefore, this fault handling approach has only a limited flexibility in regard to removing or adding plant components such as drives.

In variant III, every plant component is assigned a unique component ID. This component ID is not only used in the software, but also in the plant's construction plan or the circuit diagram and, therefore, supports working in an interdisciplinary development team. The software includes a data block for every actuator, which contains all variables belonging to this actuator (e.g. input and output variables or internal variables). Thereby the data blocks can be mapped to the according actuator by means of the component ID. Once a fault is detected on the lowest level (in an actuator), a variable in the according data block is set. The fault variables of each actuator are collected in a super ordinate status data block (cp. Figure 5, right: Status DB of a Conveying-Module including various drives and their component IDs) and the information in this status data block is transferred to the HMI including the component IDs. The flexibility of this fault handling approach is high due to the use of component IDs—unlike variant II the actuators are not assigned a specific address in the status data block, but can be identified based on the component ID. Furthermore, the component ID associated to a fault message serves to locate the fault within the plant.

Proposed Metrics for Adaptivity Focusing on Fault Handling

To ensure the quality of service of Industry 4.0 systems, it is essential to fulfill required real-time capabilities as well as adaptability, e.g. self-configuration as well as restart after a fault.

Therefore, in the following the performance metrics according to Raibulet et al. [31] are adapted for the field of automation with focus on fault detection in real-time, since the adaption in an event of fault during runtime is mandatory in order to render the required quality of service in this respect as well. In the following we regard real-time techniques, which are able to identify faults, isolate faults and introduce countermeasures during runtime.

Initially, metrics for the evaluation of real-time behavior during identification, isolation and elimination of faults are considered and, subsequently, selected metrics regarding the proportion of detectable faults in respect to potential faults are discussed under the heading of fault coverage. The proposed metrics to measure adaptivity of aPS are depicted in Figure 6.

Real-Time Capability

Programmable Logic Controllers (PLCs), which are used in manufacturing automation, operate cyclically.

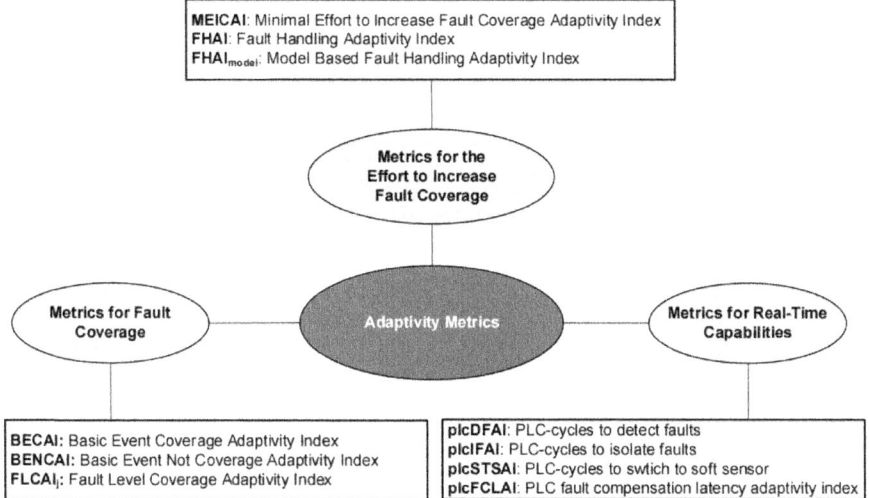

Figure 6: Proposed metrics for adaptivity for aPS.

Therefore, the number of PLC-cycles for fault detection, fault isolation and fault reaction need to be observed. Fault isolation refers to the unambiguous allocation of a fault to its cause, for example a device.

In the following a measure for implemented faults in a machine or plant needs to be found, which displays the coverage of potential faults by an automatic reconfiguration in relation to all potential faults.

$$plcDFAI = p_{cycles} \cdot t_{cycles} \tag{1}$$

An index for the survey of the punctuality of fault detection is the plcDFAI (PLC-cycles to detect faults adaptivity index, cp. Formula (1)). It is calculated by multiplying the number of cycles, which a PLC-implementation needs to detect a fault (p_{cycles}), and the cycle time (t_{cycles}), which is configured in the PLC. The result is an approximation of fault detection time based on the PLC cycle.

$$plcIFAI = i_{cycles} \cdot t_{cycles} \tag{2}$$

As soon as a fault is detected, it has to be isolated, whereas here again the fault isolation is to be understood as unit-afflicted time. Analogous to plcDFAI, the plcIFAI (PLC-cycles isolate fault adaptivity index, cp. formula (1.2)) is calculated by multiplying the number of PLC-cycles needed for fault isolation (i_{cycles}) and the configured cycle time (t_{cycles}).

$$plcSTSAI = n_{cycles} \cdot t_{cycles}$$

(3)

Once the fault is detected and isolated, strategies can be performed PLC-based in order to compensate the fault. Depending on the applied fault compensation strategy the number of PLC-cycles needed for fault compensation varies drastically. To conduct an assessment in case the method based on Schütz et al. [17] is used, the plcFCLAI (Plc-cycles for switching to software based sensor, cp. Formula (3)) is defined as the number of PLC-cycles needed to switch from real sensor to virtual sensor (n_{cycles}) multiplied with the configured cycle time (t_{cycles}).

$$plcFCLAI = plcDFAI + plcIFAI + plcSTSAI$$

(4)

The sum of the time for fault detection, fault isolation and fault compensation results in the total time for fault handling, given in Formula (4).

Real-Time Capability

The development of fault models is frequently conducted with Fault Tree Analyses (FTA) [33] in order to be able to calculate the failure probability of functions and of the overall system based on basic events and their failure rates. The term fault coverage is especially used in the field of testing [34]. According to the fault classification by Friedrich et al. [34] we examine faults that are repairable without manual intervention.

In contrast to the standardized Fault Tree Analysis, in our approach basic events are used to display physical phenomena, which need to be made measurable in order to detect the faults modeled on a superordinate level. Therefore the new term Fault coverage analysis (FCA) is used. A hierarchy of these higher-level faults can be developed in accordance with DIN 25419 [33].

By modeling the basic events the system developer is able to estimate the covered faults at an early stage of the plant's life cycle by utilizing metrics. By comparing basic events which have already been made measurable by a plant expansion stage (with sensors), fault coverage can be calculated by using the BECAI (Basic Event Coverage Adaptivity Index, cp. Formula (5)).

$$BECAI = \frac{\sum_{j=1}^{k} c_j}{\sum_{i=1}^{n} b_i} \in [0,1]$$

(5)

The BECAI corresponds to the number of basic events, which are covered by measurement-technology (c), divided by the total number of modeled basic events (b) and indicates the proportional degree of coverage ranging from 0 (0%) to 1 (100%)

Consequently, from the BECAI follows the proportion of basic events that are not covered by measurement-technology and are, therefore, not detectable. This index is called BENCAI (Basic Events Not Covered Adaptivity Index, cp. Formula (6)).

$$BENCAI = 1 - BECAI = \frac{\sum_{j=1}^{k} nc_j}{\sum_{i=1}^{n} b_i}$$

(6)

The BENCAI directly results from the proportion of basic events not covered by the BECAI or as a result of dividing the summation of all basic events which are not covered (nc) by the total amount of basic events modeled in FCA (b).

The modeled basic events in level 1 of the FCA are directly linked to the faults on level 2 (see Figure 8). If all basic events that lead to a fault are measurable, then the fault is detectable. The FLCAI (Fault Level Coverage Adaptivity Index (cp. Formula (7)) describes this coverage.

$$FLCAI_j = \frac{cf_j}{ef_j} = \frac{\sum_{n=1}^{k} cf_{(n,j)}}{\sum_{i=1}^{k} ef_{(i,j)}}; \ FLCAI_j \in [0,1]$$

(7)

Formula (7) calculates the proportion of faults which can be covered on a level j in the FCA with regard to the total number of faults modeled on level j of the FCA. This procedure is applicable to all levels of the FCA that are above the basic event level, all the way up to the top-level node of the FCA. The result is the fault coverage on level j in percent.

Minimal Programming Effort to Increase Fault Coverage

The fault coverage formula in Formula (7) also indicates the number of uncovered faults. In order to examine, which fault may be detected with a minimal enhancement effort, the MEICAI (Minimal Effort to Increase Fault Coverage Adaptivity Index, Formula (8)) is defined.

$$MEICAI_j = \min\left(\bigcup_{i=1}^{n} fault_{(i,j)} \right);$$

$$fault_{(i,j)} = \frac{\sum BENC_{(i,j)}}{\sum BEC_{(i,j)}}; \quad 0 < MEICAI_j \leq 1$$

(8)

The MEICAI shows those faults i on FCA level j with the highest number of coverage of basic events covered (BEC) in proportion to basic events not covered (BENC). All faults on level j are included and the above-mentioned calculation is carried out, whereby the fault is chosen, which has the minimum aspect ratio of BENC to BEC and, therefore, contains few BENC but a maximum amount of BEC. Only faults with a number of BENCs higher than 0 are included in this calculation because otherwise the MEICAI$_j$ is 0 and, therefore, all basic events are already measurable and covered.

The MEICAI$_j$ indicates the fault that can be covered with minimum effort. In order to estimate the required effort further, for mechatronic systems changes of software and automation hardware as well as changes of the mechanics need to be considered. This article focuses on software changes.

In order to implement the identified fault detection and compensation strategies, with minimum effort for fault compensation in software, adaptations in the software is needed that have to be made in the languages of the IEC 61131-3 standard-in this example the language function block diagram is chosen. The internal behavior of the FBDs is thereby not considered (cp. Fuchs et al. [35]).

For an estimation of the effort for the required software changes the FHAI (Fault Handling Adaptivity Index, cp. Formula (9)) is introduced.

$$FHAI = \frac{FHAI_{new} + FHAI_{adapted} + FHAI_{removed}}{FHAI_{new} + FHAI_{adapted} + FHAI_{removed} + FHAI_{old}} \in [0,1]$$

(9)

The FHAI calculates the proportion of newly added ($FHAI_{new}$), adapted ($FHAI_{adapted}$), removed ($FHAI_{removed}$) and unchanged ($FHAI_{old}$) software elements of each function block in IEC 61131-3. If, for example, all software elements have been changed and, thus, no unchanged software elements remain, a proportionate effort of 1 (100%) needs to be summoned because $FHAI_{old}$ equals 0; whereas, if no changes are conducted, the FHAI is 0. Therefore, the FHAI gives the proportion of software changes, which need to be conducted in order to guarantee adaptivity, in percent.

The single factors of FHAI are defined as follows within (10) to (13).

$$FHAI_{new} = \sum i_{new} + \sum o_{new} + \sum fb_{new};$$
$$i_{new}, o_{new}, fb_{new} \in \mathbb{N}_0^+$$

$$(10)$$

$$FHAI_{adapted} = \sum i_{adapted} + \sum o_{adapted} + \sum fb_{adapted};$$
$$i_{adapted}, o_{adapted}, fb_{adapted} \in \mathbb{N}_0^+$$

$$(11)$$

$$FHAI_{removed} = \sum i_{removed} + \sum o_{removed} + \sum fb_{removed};$$
$$i_{removed}, o_{removed}, fb_{removed} \in \mathbb{N}_0^+$$

$$(12)$$

$$FHAI_{old} = \sum i_{old} + \sum o_{old} + \sum fb_{old};$$
$$i_{old}, o_{old}, fb_{old} \in \mathbb{N}_0^+$$

$$(13)$$

For instance, $FHAI_{new}$, Formula (10), is calculated by summarizing the sum of newly added inputs (i_{new}) and newly added outputs of the function block (o_{new}) and the sum of newly added function blocks itself (fb_{new}). The calculations for adapted ($FHAI_{adapted}$), removed ($FHAI_{removed}$) and retained software elements ($FHAI_{old}$) are carried out equivalently. Using IEC 61131-3 function block diagram, it may occur that the output of a function block is, at the same time, the input for one or several other function blocks. These interconnections are included in the FHAI calculations as inputs as well as outputs, since they either remain unchanged as inputs and outputs (o_{old}) or they are changed (o_{new}, $o_{adapted}$ and $o_{removed}$).

In this section the programming effort was measured. In Schütz et al. [19] the necessary code was generated from a model (SysML), therefore for an MDE based approach instead of a programming effort the effort to adapt the model should be evaluated. But nevertheless for the maintenance staff the additional programming effort is interesting as an indirect measure program comprehension according to Vogel-Heuser [36] .

Minimal Modeling Effort to Increase Fault Coverage

Frey and Litz [37] introduced complexity metrics for Petrinets using besides others an adapted McCabe metric. Chidamber and Kemerer developed a set of metrics of OO design [38] , e.g. weighted methods per class (WMC) which is a measure for class complexity used in this paper: in order to calculate the WMC of a program, the cyclomatic complexity measure of each method is summed up for all classes, cp. [39] . WMC and the cyclomatic complexity measure have already been successfully applied comparing task complexity for usability evaluation [36] . The cyclomatic complexity measure by McCabe [40] is calculated as stated in Formula (14),

$$V(G) = e - n + 2 \cdot p \tag{14}$$

with e being the number of edges (like transitions in a state chart), n being the number of nodes (states of a state chart) and p being the number of connected components (number of analyzed state charts). The complexity measure V(G) represents the number of linearly independent paths through the analyzed graph.

In the case of model changes in UML or SysML state charts the McCabe metric is applied instead of the metrics introduced in Section 4.3, since the necessary changes of the state charts are restricted to adding or removing states and transitions.

As a metrics for the relative modelling effort $FHAI_{model}$:

$$FHAI_{model} = \frac{\left(V(G)_{new} - V(G)_{old}\right)}{0.5 \cdot \left(V(G)_{new} + V(G)_{old}\right)} \tag{15}$$

This mathematical comparison can also be used to compare the complexity of software evolutions, standardized by the mean value of compared software evolution. Due to the standardization, it is possible to compare pairwise different independent software evolutions, to point out an evolution, in which complexity does increase too much, e.g. to ease software maintenance.

Furthermore, three cases concerning $FHAI_{model}$ can be stated shown in Formula (16). The first case indicating that the cyclomatic complexity measure of both models stages is equal. Hence, $FHAI_{model}$ indicates no increasing or decreasing model complexity. Furthermore, as soon as the new model evolution has a bigger cyclomatic complexity, $FHAI_{model}$ indicates it resulting in a positive number. The last case indicating a decreasing model complexity in case of the new cyclomatic complexity is less than the old cyclomatic complexity, resulting in a negative number.

$$V(G)_{new} = V(G)_{old} \rightarrow FHAI_{model} = 0$$
$$V(G)_{new} > V(G)_{old} \rightarrow FHAI_{model} > 0$$
$$V(G)_{new} < V(G)_{old} \rightarrow FHAI_{model} < 0 \tag{16}$$

Evaluation of Defined Metrics by Means of Three Different Types of Adaptive Strategies Using Two Different Laboratory Demonstrators

In this section the metrics defined in the previous Section 4 are applied to different simple lab demonstrators: a pick and place unit plant (a) PPU (cp. Figure 7) as well as its extended version (b) (cp. Figure 16 in Section 5.4) (both demonstrators of the PP 1593 design for future-managed software evolution) and a tank of a process plant (c) (cp. Figure 14).

For all four scenarios the realized concept of adaptation is shortly discussed at first and thereafter the scenario is evaluated using the proposed metrics.

Using two different lab size demonstrators four different application examples representing different types of adaptivity are introduced in this section. The first and the third application scenario represent fault detection, fault isolation of a faulty device, e.g. a sensor and self-configuration to continue operation. The second and fourth represent strategies to restart after a malfunction, e.g. a stop or emergency shutdown. In the first application example sensor redundancy is given and on this basis the faults can be isolated (5.1) as a prerequisite for later restart. Self-configuration of a model based redundancy for a tank level in case of a sensor fault is discussed in Section 5.3. The second scenario stores the state of the machine in case of an emergency shut-down and remembers the position of the work piece to restart properly. The fourth application example uses the extended xPPU for a state based restart based on an OMAC state machine in case of faulty position of work piece and a resulting machine stop (5.4).

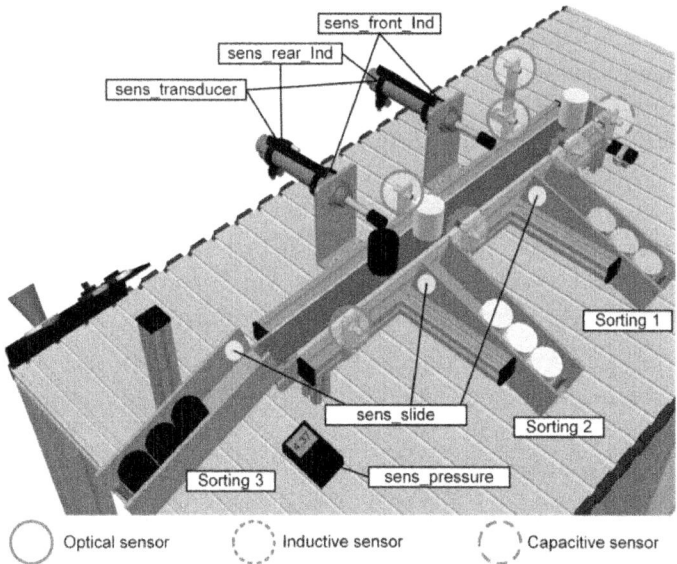

Figure 7: Converyor belt of the lab size application example Pick and Place Unit (PPU).

PPU-FCA Based Fault Detection and Self-Configuration with Redundant Sensor Device

In the sorting unit of the plant work pieces are distributed to three different slides (Sorting 1-3) according to their color and material. Each slide is equipped with an inductive sensor (sens_slide). Each pusher has a front and a rear end position sensor as well as an analogous sensor for precise position measurement. The analogue sensor represents the redundant sensor device to measure the position of the pusher. All malfunctions leading to the report of a timeout error and the malfunctions' causes are considered (cp. Figure 8) and it is analyzed to what extent it is possible to automatically assign them.

In Figure 8, a fault tree with potential causes (basic events) and their logical contexts, those indicate a fault, are depicted. The fault tree itself only refers to the conveyor unit (cp. Figure 7) consisting of sensors for material detection in order to sort different kinds of work pieces in according slides. The light gray colors indicate causes, which can be detected with a basic configuration of the PPU via measurement technology. For instance, basic event p7 represents a binary sensor at the slide of the sorting unit, which detects the incoming material of a work piece.

To identify failures to be handled automatically, the FCA is used. The realization of self-configuration functionality presumes that failures have to be observable by the software (i.e. adequate sensors are installed to identify a specific failure) and automatic compensation mechanisms (analogously to maintenance instructions) have to be implemented. Therefore, the PPU's self-configuration variant consists of additional sensors. For example, to detect a "WP Jam" within the sorting station, pushers previously consisting of binary positioning sensor are substituted by pushers equipped with analogue transducers to precisely detect the pusher's positions (parameter sens_transducer in Figure 7). The installation of an additional pressure sensor facilitates the detection of pressure failures (cp. Figure 8, p. 1). Sensors are installed at each slide to detect arriving WPs by monitoring for rising edges of the sensor signal (sens_slide, cp. Figure 7).

In the following the metrics defined in Section 4 are applied to the PPU.

Metrics for Real-Time Capabilities

According to the definition in Section 4.1 various metrics for real-time have been defined. The metrics for real-time capabilities are applied to two software variants implemented in sequential function chart (SFC, defined in the IEC 61131-3 standard, cp. Figure 9). The variants are compared in order to determine (by means of the metrics) which one of the two implementations shows better real-time characteristics; thus, shorter cyclic times for plcFCLAI (cp. Formula (4)).

Fault detection, selection and compensation are performed simultaneously in three states (cp.Figure 9, left). In an IEC 61131-3-runtime environment simultaneous states are passed through within one cycle time. Thus, the plcFCLAI results to 30 ms, on the assumption that the PLC-cycle time is configured to 30 milliseconds.

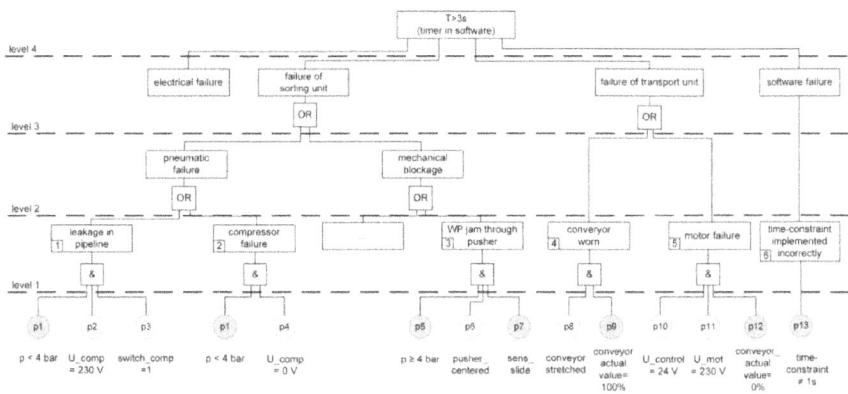

Figure 8: Part of a FCA regarding time monitoring of work piece sorting for conveyor belt of PPU.

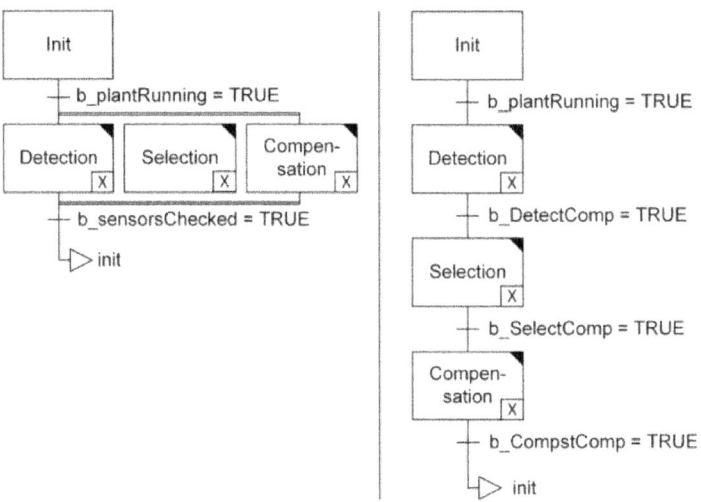

Figure 9: Different implementations for fault detection, selection and compensation.

With a sequential implementation of fault detection, selection and compensation (cp. Figure 9, right side), initially the time needed for fault detection (plcDFAI) has to be calculated according to (1.1). On the assumption that fault detection lasts one cycle (p_{cycles} = 1) and the cycle time is adjusted to 30ms (t_{cycles} = 30 ms), the plcDFAI also results in 30 ms.

The same is valid for fault selection, which is done in the next PLC-cylce. This again lasts one cycle time (i_{cycles} = 1) and the plcIFAI also results in 30 ms. After fault selection the fault compensation strategy is initiated. The initiation itself also lasts a cycle time (n_{cycles} = 1)

$$plcSTSAI = n_{cycles} \cdot t_{cycles} = 1 \times 30 \text{ ms} = 30 \text{ ms}$$

The index for total real-time capability in the event of a fault (plcFCLAI) is then calculated with the single factors for real-time capability

$$plcFCLAI = plcDFAI + plcIFAI + plcSTSAI = 30 \text{ ms} + 30 \text{ ms} + 30 \text{ ms} = 90 \text{ ms}$$

By applying the metric for real-time capability to two different software implementations it was shown that the first variation (plcFCLAI = 30 ms) holds better real-time characteristics than the second implementation (plcFCLAI = 90 ms).

Metrics for Fault Coverage

In this section the metrics for fault coverage defined in Section 4.2 are applied to the introduced fault tree (cp. Figure 8) of the conveyor unit. At first the index BECAI (Basic Event Covered Adaptivity Index) is calculated according to (6). For this only the existing basic events of the fault tree are regarded. It is checked which basic events are measurable by the current PPU-configuration (factor c). This factor is divided by the total amount of existing basic events b. Thus, the BECAI results in:

$$BECAI = \frac{\sum_{j=1}^{k} c_j}{\sum_{i=1}^{n} b_i} = \frac{7}{14} = 0.5 \rightarrow 50\%$$

With the current configuration of the system, 50% of the basic events are measurable.

The BENCAI (Basic Events Not Covered Adaptivity Index) amounts to 0.5 (50%), since 7 out of 14 basic events are not recordable by measurement techniques.

$$BENCAI = 1 - BECAI = 0.5 = \frac{\sum_{j=1}^{k} nc_j}{\sum_{i=1}^{n} b_i} = \frac{7}{14}$$

Since faults can be detected by the logical connection of basic events, the faults, that are detectable by registration with measurement techniques, are calculated. For this the index FLCAI$_j$ (Fault Level Coverage Adaptivity Index) for the level next above the basic events (j = 2) is determined. It is, therefore, necessary, to calculate how many faults are detectable based on the measurement technology. Applied to the fault tree, FLCAI2 (Formula (7)) results in:

$$FLCAI_2 = \frac{\sum_{n=1}^{k} cf_{(n,2)}}{\sum_{i=1}^{k} ef_{(i,2)}} = \frac{1}{6} = 0.167$$

Although 50% of the basic events are measurable, within the current configuration only one fault modeled in the fault tree is detectable, since only for this one fault (time constraint incorrect) all connected basic events are recordable. In order to increase the fault coverage it is necessary to adapt software and automation hardware as well as the mechanics of the PPU.

Metrics for the Programming Effort Needed to Increase Fault Coverage

In order to estimate the minimum effort to increase fault coverage, or rather to detect a fault, which can be turned into a measurable fault with a minimum amount of effort, the metrics defined in Section 4.3 are applied to the conveyor unit.

Initially, the fault modeled in the fault tree, which already has a very high amount of measurable basic events, is determined. For this the index MEICAI2 is applied. For all modeled faults on level 2 (j = 2), the amount of non-measurable basic events is divided by the total amount of modeled basic events. Subsequently, the value with the minimum ratio is selected. According to Formula (8) this leads to the following calculation:

At first the ratios are calculated for each fault on level 2.

$$\left(\begin{pmatrix} \text{fault}_{(1,2)} = \frac{\sum \text{BENC}_{(1,2)}}{\sum \text{BEC}_{(1,2)}} = \frac{2}{3} = 0.67 \end{pmatrix}, \begin{pmatrix} \text{fault}_{(2,2)} = \frac{\sum \text{BENC}_{(2,2)}}{\sum \text{BEC}_{(2,2)}} = \frac{1}{2} = 0.5 \end{pmatrix},\right.$$

$$\begin{pmatrix} \text{fault}_{(3,2)} = \frac{\sum \text{BENC}_{(3,2)}}{\sum \text{BEC}_{(3,2)}} = \frac{1}{3} = 0.33 \end{pmatrix}, \begin{pmatrix} \text{fault}_{(4,2)} = \frac{\sum \text{BENC}_{(4,2)}}{\sum \text{BEC}_{(4,2)}} = \frac{1}{2} = 0.5 \end{pmatrix},$$

$$\left.\begin{pmatrix} \text{fault}_{(5,2)} = \frac{\sum \text{BENC}_{(5,2)}}{\sum \text{BEC}_{(5,2)}} = \frac{2}{3} = 0.67 \end{pmatrix}, \begin{pmatrix} \text{fault}_{(6,2)} = \frac{\sum \text{BENC}_{(6,2)}}{\sum \text{BEC}_{(6,2)}} = \frac{0}{1} = 0 \end{pmatrix},\right)$$

Formula (1.8) requires the minimum ratio, whereby those MEICAI2, which equal zero, are excluded from further calculations. A zero is equivalent to a 100% metrological coverage of the basic events (cp. fault (6, 2)). Thus, the minimum MEICAI is fault (5, 2) with MEICAI2 = 0.33; which in the fault tree (cp. Figure 8) corresponds to a work piece jam through the pusher.

The effort for an extension of the software is expressed by the FHAI (cp. Formula (9)). In the function block diagram of the PPU (cp. Figure 10) inputs, outputs and function blocks have been retained as well as added or adapted. None of the software elements have been removed. Therefore, in this example only FHAI_{new}, $\text{FHAI}_{adapted}$ and FHAI_{old} (Formulas (10), (11), and (13)) are applied in order to calculate the FHAI.

Adaptations to FB_monitoring

FB_Monitoring in Figure 10 illustrates that inputs, outputs and function blocks have not been changed. Thus, only FHAI_{old} has to be calculated which results in:

$$\text{FHAI}_{old} = \sum i_{old} + \sum o_{old} + \sum fb_{old} = 3 + 1 + 1 = 5$$

With i = 3 referring to the three inputs and o = 1 in accordance with the single output. The internal structure of FB_Monitoring has not been changed (fb = 1).

Adaptations to FB_monitoring_adaptive

The whole function block FB_Monitoring_adaptive has been newly added in order to process the new measured values and transfer the according fault status to FB_Fault_Hanlding.

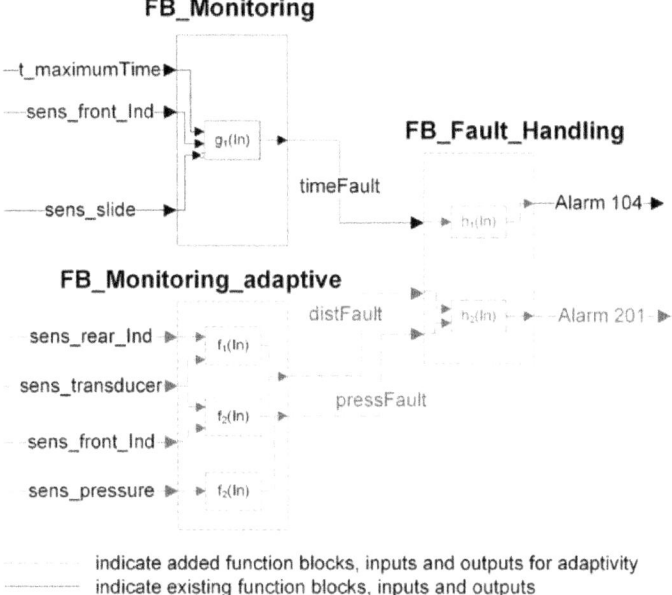

indicate added function blocks, inputs and outputs for adaptivity
indicate existing function blocks, inputs and outputs

Figure 10: Software adaptation from the basic PPU configuration for detecting the fault work piece jam using self-configuration.

Thus, for this function block $FHAI_{new}$ needs to be calculated. This results from the four newly added inputs ($i = 4$), two newly added outputs ($o = 2$) and the new function block FB_Monitoring_adaptive itself ($fb = 1$).

$$FHAI_{new} = \sum i_{new} + \sum o_{new} + \sum fb_{new} = 4 + 2 + 1 = 7$$

Adaptations to FB_fault_handling

FB_Fault_Handling includes new, adapted and also retained software elements. $FAHI_{old}$ results from the retained input ($i = 1$) and the retained output ($o = 1$). Thus it is calculated to

$$FHAI_{old} = \sum i_{old} + \sum o_{old} + \sum fb_{old} = 1 + 1 + 0 = 2$$

The newly added software elements ($FHAI_{new}$) are, for once, the new inputs of the function block ($i = 2$) and also the newly added output ($i = 1$). The $FHAI_{new}$ results in

$$FHAI_{new} = \sum i_{new} + \sum o_{new} + \sum fb_{new} = 2 + 1 + 0 = 3$$

Since the internal structure of FB_Fault_Handling has been adapted, $FHAI_{adapted}$ results in

$$FHAI_{adapted} = \sum i_{adapted} + \sum o_{adapted} + \sum fb_{adapted} = 0 + 0 + 1 = 1$$

Overall effort of the software changes (FHAI)

Based on the previous calculations, it is possible to sum up the individual indexes $FHAI_{old}$, $FHAI_{new}$ und $FHAI_{adapted}$ in order to calculate the overall effort of software adaptations (FHAI).

$FHAI_{old} = 2 + 5 = 7$

$FHAI_{new} = 3 + 7 = 10$

$FHAI_{adapted} = 1$

Thus the overall effort of software adaptations is calculated according to Formula (9) and results in

$$FHAI = \frac{FHAI_{new} + FHAI_{adapted} + FHAI_{removed}}{FHAI_{new} + FHAI_{adapted} + FHAI_{removed} + FHAI_{old}}$$

$$= \frac{10 + 1 + 0}{10 + 1 + 0 + 7} = \frac{11}{18} = 0.61$$

The FHAI is equal to 0.61 which means that overall 61% of the software elements need to be added or adapted. Thereby, internal structures such as the addition or adaptation of internal function blocks in Figure 10 are not included and would lead to an even higher overall effort.

Metrics for the Modeling Effort Needed to Increase Fault Coverage

In Figure 11, an excerpt of the PLC software based on plcUML [41] is given. In the manual mode (left hand side), the software is not able to identify a WP jam precisely and therefore triggers an alarm (201)

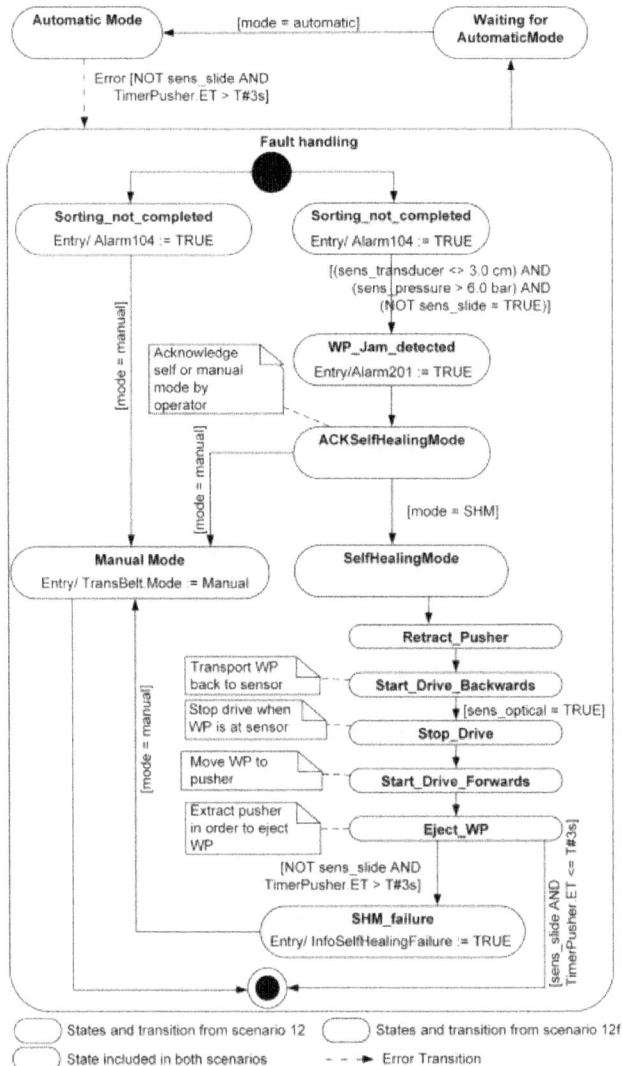

Figure 11: Excerpt of plcUML state chart for manual and self-healing mode [16] .

if a product does not reach the slide within the expected time (Timer t > 3 seconds). Obviously, this is a drastically simplified failure model. In contrast, the additionally installed sensors in the self-configuration case provide the necessary information to detect a WP jam more precisely. This failure can be exactly determined when

- the transducer contained within the respective pusher indicates that it is not extended completely (sens_transducer),

- the pressure of the pneumatic system is not as expected (sens_pressure) and
- no WP entered the slide (sens_slide).

According to [42] , automatic and manual mode is recommended in P&M. The self-configuration software functionality is realized as an additional functionality of the automatic operation mode. After detecting the fault and the corresponding failure, a sequence of operations for fault recovery, i.e. to remove the WP jam and the correct sorting of the WP, is executed (cp. Figure 11).

To detect the complexity of both plcUML state charts $FHAI_{model}$ is applied. To calculate $V(G)_{old}$ all states are taken into account used in scenario 12 (cp. Figure 11, white, white-gray colored and initial and end state). Furthermore, all edges are counted that are related to a state from scenario 12. Resulting, 6 states and 6 transitions are realized for scenario 12. Hence, $V(G)_{old}$ is 2 regarding Formula (14).

$$V(G)_{old} = e - n + 2 \cdot p = 6 - 6 + 2 \times 1 = 2$$

An evolutionary plcUML implementation is also shown in Figure 11 (gray, white-gray colored and initial and end state). The new implementation is realizing the more precise fault detection, as mentioned above. There are 16 transitions and 19 states, resulting in a cyclometic complexity of 5

$$V(G)_{new} = e - n + 2 \cdot p = 19 - 16 + 2 \times 1 = 5$$

Using both results, $FHAI_{model}$ can be calculated (Formula (15)) to get the relative complexity gradient as follows:

$$FHAI_{model} = \frac{\left(V(G)_{new} - V(G)_{old}\right)}{0.5 \cdot \left(V(G)_{new} + V(G)_{old}\right)} = \frac{5-2}{0.5 \times (5+2)} = \frac{3}{3.5} = 0.857$$

$FHAI_{model}$ reveals to be about 0.857 indicating increasing complexity of the new plcUML implementation concerning the previous software evolution.

Restart of PPU after Emergency Shut-Down

In another part of the PPU shown in Figure 16, workpieces get stamped using two pneumatic cylinders. One of the pneumatic cylinders pulls the workpieces in to reach stamp position, one cylinder with adjustable pressure stamps the workpieces. In this case, the PPU is programmed using the adapted UML state charts for PLCs integrated within the CODESYS environment, which is introduced and explained in [41] (cp. Figure 12).

In a scenario during testing of the operation of the PPU the emergency stop is activated during stamping. When restarting after the emergency stop the PPU initializes its modules anew and starts operation beginning from a stack, where the workpieces are stored. When trying to stamp a workpiece the stamp has the old workpiece still in place, which collides with the new workpiece when the PPU tries to set the workpiece down. In order to eliminate this behavior a Failure Mode and Effects Analysis (FMEA) is conducted. An excerpt of the FMEA is shown in Figure 13. It is found that after initialization no analysis of the current state of the PPU and especially no analysis of workpieces that may still be within the PPU is done. Consequently, as a prevention, new conditions and more specifically transitions and one new state to firstly analyze the state of the PPU after start and restart and its workpieces is introduced.

In order to measure the effort required to implement identification of work piece positions after an emergency stop of the PPU, the complexity measure cyclomatic complexity measure V(G) by McCabe [40] (introduced in Section 4.4) is used. In order to implement the storing of work piece positions, only the state chart controlling the crane needs to be adapted. Depending on the position of a work piece after emergency stop, the crane needs to turn to the according station to resume the interrupted transport. The original state chart controlling the crane consists of 17 edges (e = 17), 15 states (n = 15) and only one state chart is needed to control the crane (p = 1). Thus, the V(G)$_{old}$ results in (cp. Formula (14)

$$V(G)_{old} = 17 - 15 + 2 \times 1 = 4$$

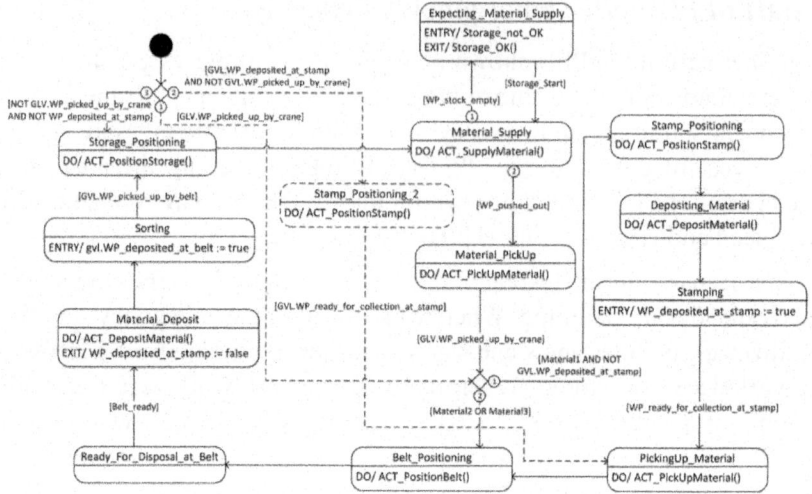

Figure 12: PPU implementation of restart after emergency shut down as UML state chart implemented in CoDeSys V3 (dashed lines indicate added edges and elements).

FMEA-form sheet for the system "stamping plant"

Function: Resume operation after emergency stop		
Potential failure sequences	Potential causes of failure	Prevention
Unknown position of material in plant	Undetected material in plant after emergency stop	Save momenrtary state of machine and material when emergency stop is initiated
> Material in stamp is forgotten		Design program more flexibly according to actual state of plant
> Material on crane is not transported to destination		
>> Unknown behavior of plant and material	Unknown behavior after initialization after emergency stop	Restart only after all material is removed or position of material is considered in program
>>> Safety risk for operators and machine		

Figure 13: Failure mode and effects analysis for collision of workpieces within stamp.

To enable the storing of work piece positions, an additional state as well as three additional transitions are necessary resulting in a state chart with a total of 20 transitions (e = 20) and 16 states (n = 16), while the number of analyzed state charts remains unchanged (p = 1). Therefore, the cyclomatic complexity measure $V(G)_{new}$ of the expanded crane state chart, according to Formula (14), calculates to

$$V(G)_{new} = 20 - 16 + 2 \times 1 = 6$$

Overall, the cyclomatic complexity measure of the state chart increases by two when implementing the memorizing of work piece positions, which means (regarding Formula (15)) that the relative complexity measure ($FHAI_{model}$) increases by 0.4.

$$FHAI_{model} = \frac{V(G)_{new} - V(G)_{old}}{0.5 \cdot (V(G)_{new} + V(G)_{old})} = \frac{6 - 4}{0.5 \times (6 + 4)} = 0.4$$

Self-Configuration with Model Based Redundancy for Liquid Level in Tank

For another example a tank, equipped with two sensors detecting the tank's filling level (maximum and minimum), is considered. The tank is used to store liquids in a production process (cp. Figure 14). Once the minimum filling level of the tank is reached (detected by the lower tank sensor, sensor 101.2 in Figure 14), a valve (V101) is opened and the filling process starts. The filling process will run until the upper tank sensor (sensor 101.1) signals that the maximum filling level of the tank is reached. However, there is no redundancy in the process and, thus, a malfunction of the upper sensor leads to an overflow of the tank. As a precaution and in order to detect a failure of the upper sensor, the original tank function is enlarged by a self-configuration method according to Gronau's classification of adaptivity [29]. For self-configuration the filling level of the tank is calculated based on the amount of fluid that is filled into the tank, the amount of fluid that flows out of the tank, the tank's dimensions as well as the filling and emptying times. In the following, the introduced metrics to estimate the programming effort (cp. Section 4.3) are used to measure the effort of implementing this function on code level.

A transformation of the approach presented by Schütz et al. [17] into a fault coverage analysis tree (FCA) is presented in Figure 15. On the bottom of the FCA the basic events connected to a fault are shown, i.e. if inflow or outflow or the level sensor is faulty, a filling level error and an inflow error and an outflow error would occur.

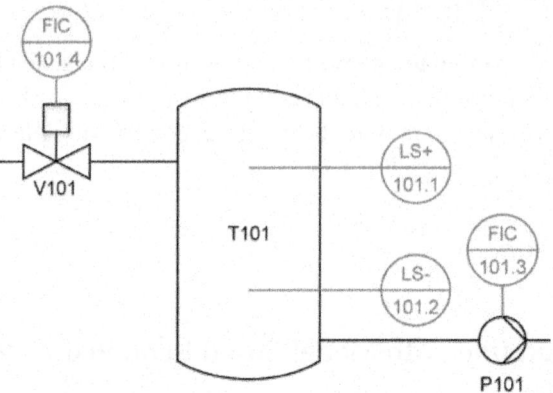

Figure 14: Tank with upper and lower filling level sensors, valve and pump.

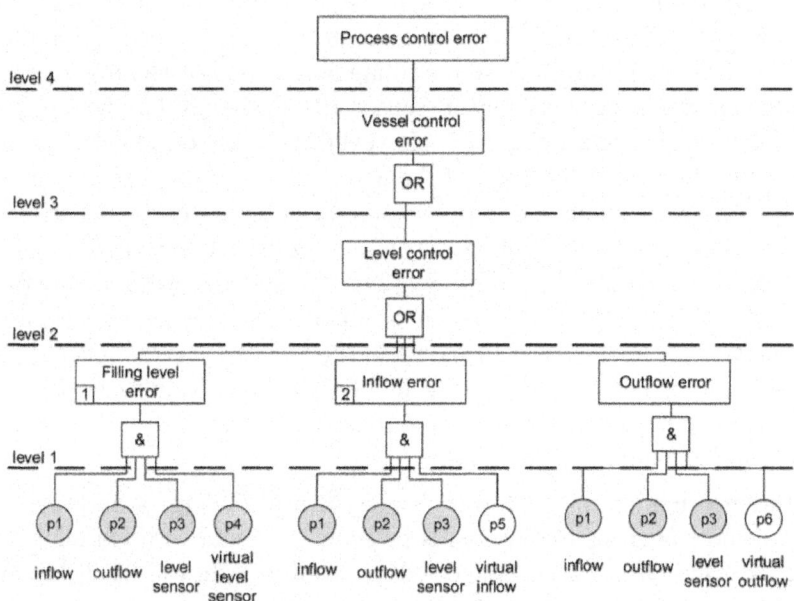

Figure 15: Part of a FCA regarding tank's level control error.

As discussed above the real level sensor can be approximated by a virtual sensor (e.g.Figure 15, virtual filling level sensor) that is calculated during runtime and is added as basic event.

If the filling level sensor is effected by a fault as mentioned above, the faulty situation may be detected, but the reason for it remains unclear compared to the FCA in Figure 4 case study 5.1, e.g. the broken cable (electric fault), work piece jam (mechanical fault) or wrong software implementation (faulty software design). At first the redundancy model has to be developed in the design phase by taking functional dependencies into account formulated as physical equation. The redundancy model has to be implemented on the PLC to allow decisions during runtime. But nevertheless the approach allows only the self-configuration of one fault out of three real sensors and one calculated virtual one. BECAI may be calculated to 100% for filling level error neglecting physical reasons for such a fault, e.g. faulty cable, voltage, input channel etc. For level control error ($j = 3$), one out of three will be detected, resulting in:

$$FLCAI_3 = \frac{1}{3} = 0.\overline{333}$$

Metrics for Real-Time Capabilities

In order to detect a malfunction of the upper sensor, no additional PLC cycle is necessary, since the filling level of the tank is calculated simultaneously to the normal program execution. Therefore, only cycle time is needed to detect a sensor failure (plcDFAI) and no additional time is needed to select (plcIFAI) and compensate (plcSTSAI) the calculated value compared to reacting to the actual sensor value, since the actual sensor value and the calculated value are checked at the same time (as two options in the same line of code). Thus, the metrics for real-time capabilities result in

$$plcDFAI = 1 \cdot t_{cycles} = t_{cycles}$$

$$plcIFAI = plcSTSAI = 0$$

Overall, the index for total real-time capability in the event of a malfunction of the upper tank sensor calculates to

$$plcFCLAI = plcDFAI + plcIFAI + plcSTAI = t_{cycles} + 0 + 0 = t_{cycles}$$

Metrics for the Programming Effort Needed to Increase Fault Coverage

The original declaration part of the tank function (without self-configuration) consists of 14 lines of code (LOC) and the original implementation part consists of 22 LOC (not including comments). The whole declaration part and 21 LOC of the implementation part are retained in the new function (with self-configuration) and, thus, the $FAHI_{old}$(calculates the unchanged software elements) adds up to

$$FHAI_{old} = 14 + 21 = 35$$

In order to implement self-healing, neither in the declaration part nor in the implementation part of the function LOC are removed ($FHAI_{removed} = 0$). Furthermore, in the declaration part of the self-healing function 8 LOC were added and in the implementation part 7 LOC were added. Therefore, the $FHAI_{new}$ is calculated as follows

$$FHAI_{new} = 8 + 7 = 15$$

Also, in the implementation part one LOC was modified to implement self-healing in the function ($FHAI_{adapted} = 1$). Overall, the FHAI results to

$$FHAI = \frac{FHAI_{new} + FHAI_{adapted} + FHAI_{removed}}{FHAI_{new} + FHAI_{adapted} + FHAI_{removed} + FHAI_{old}}$$

$$= \frac{15 + 1 + 0}{15 + 1 + 0 + 35} = \frac{16}{51} = 0.31$$

This means in order to enlarge the function of the tank and thus detect and cover a failure of the upper tank filling sensor, 31 percent of the software code had to be modified or added.

Evaluation of Enhanced Recovery States Based on OMAC State Machine with xPPU

In order to evaluate the proposed metrics, another program (programmed by another engineer with a bachelor degree in computer science as well) with and without recovery function for the laboratory demonstrator xPPU is analyzed. The xPPU (cp. Figure 16) is an extension of the laboratory demonstrator introduced in Section 5.1. Work pieces are stored at the stack and, depending on their material, are either transported with the crane module to the stamping module before being sorted in the sorting unit (described in Section 5.1) or are directly transported to the sorting unit. Additionally to the features of the original PPU, the conveyor belt depicted in Figure 7 has been enlarged by a conveyor system, which enables re-feeding of work pieces from the original conveyor belt back into the manufacturing process and is also able to change the order of the work pieces.

The analyzed program is divided into unit modules, equipment modules and control modules according to the ISA88 physical hierarchy for code modules, which can be seen in Figure 17. The unit module is controlled by an OMAC state machine [15] (cp. Section 2 for description) thereby the active OMAC state calls the suitable function of the unit module. Subsequently, the suitable functions of the equipment modules are called by the unit module and the control modules by the associated equipment module. The return value of a function call is given as an enumeration (OPERATING, FINISHED, ERROR) which is an important part of the failure detection and recovery system. Additionally to the OMAC states and contrary to the traditional development process of field control software, the approach suggests the use of system states whereby machine capabilities are implemented as basic operations with pre- and post-conditions.

Therefore, additional auxiliary functions are needed for modeling the operations, pre- and post-conditions. To apply failure detection, the execution time of every basic operation is measured and used for the definition of a failure, which is defined as the exceedance of the average execution time of a basic operation by a chosen offset of 300ms. Thus the program of the PPU was enlarged by a recovery function, which enables the detection of a fault and transfers the plant into a recovery state.

Comparing the program described above to case study E, it is obvious that not only the architecture levels correspond to each other. Case study E likewise has got a system state given by its facility wide module control, yet with a more loose coupling to the architecture level below the module control, i.e. the modules in case study E. This allows a differentiated reaction to an error regarding the states of the modules. Nevertheless, like in the example shown in Figure 17, an occurring error in case study E also has an immediate influence on the state of its facility wide module control and thereby on the system state as the module control commands the states of all inferior modules.

In the following, the introduced metrics (cp. Section 4) are used to measure the adaptivity of the demonstrator program.

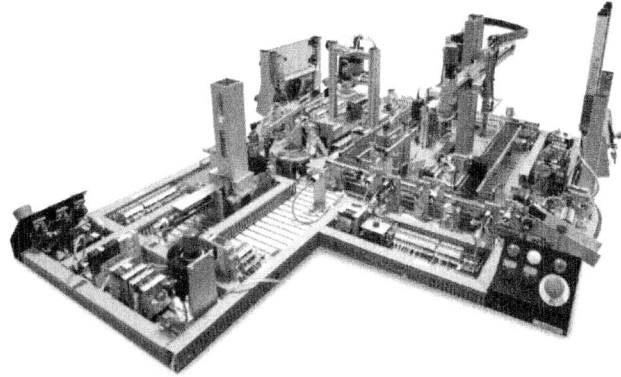

Figure 16: Picture of the extended Pick and Place Unit (xPPU).

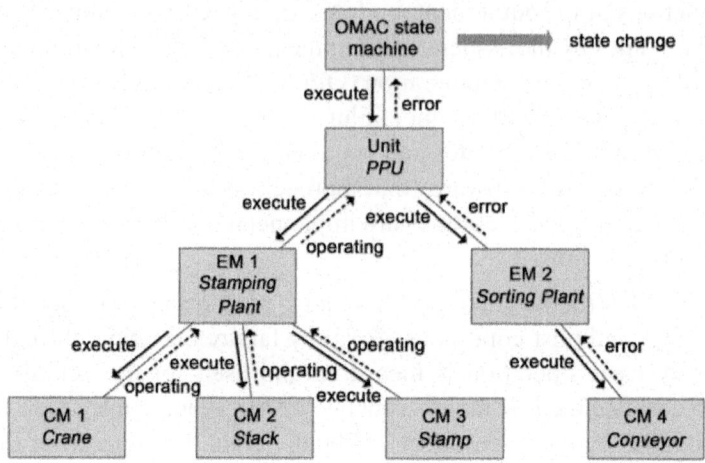

Figure 17: Fault detection and transmission cascading through the architecture levels OMAC state machine, unit, equipment module (EM) and control module (CM).

Metrics for Real-Time Capabilities

In this section the real-time capability of this approach is analyzed. After initialization, the demonstrator program is executed in the state Execute. In order to monitor the fault-free execution of the program, the average time needed for the individual operations is measured during program execution. The considered failure is a false timing of the pusher-extraction, which is used for sorting the work pieces into slides. The failure may occur due to differences in conveyor speed and, therefore, a false timing between conveyor speed and pusher-extraction. Another reason for this failure can be the different friction behavior of the different types of work pieces. The failure is detected as follows: In case of fault-free execution the work piece to be sorted is detected by the according slide sensor (cp. Figure 7 sens_slide) after extracting the pusher. If however, the described timing fault occurs, the work piece is not correctly pushed into the slide and, therefore, the sensor sens_slide does not report the detection of a work piece. After the average time for the operation sort work piece is expired, an additional delay timer of 300 ms is started. This offset of 300 ms is chosen for the entire demonstrator to ensure that any action carried out by the demonstrator is definitely finished within the delay time. Once the failure is detected in the control module level, the return value ERROR is passed through the equipment and unit module level to the OMAC state machine and leads to a state transition to the OMAC state Holding (cp. Figure 17). The offset between the time, when a failure is detected, and the average execution time leads to a value of plcDFAI = 300 ms

In this state the program execution is stopped and the plant is transferred into a safe state. Once a safe state is obtained, the state is changed to Held and, for security reasons, no further actions are performed until the machine operator starts the recovery. Changing the state of the plant from Execute to Holding requires one cycle. Stopping the plant in a safe state and changing to the state Held, which completes the fault selection, takes at most two cycles. With a cycle time of 4ms ($t_{cycles} = 4$ ms), the plcIFAI results in

$$plcIFAI = 3 \cdot t_{cycles} = 3 \times 4 \text{ ms} = 12 \text{ ms}$$

Subsequently, the fault compensation strategy is started after completion of the fault selection. After an unknown time period $t_{operator}$, which passes between fault selection and the operator's user input to start recovery, the plant is transferred to the state Un-Holding in which automatic recovery is started. Once the operator's input is detected, the initiation of the fault compensation is started either in the same cycle as the input or in the following cycle leading to a maximum time of two cycle times. In consideration of the unknown time period $t_{operator}$, the plcSTSAI calculates to

$$plcSTSAI = 2 \cdot t_{cycles} + t_{operator} = 2 \times 4 \text{ ms} + t_{operator} = 8 \text{ ms} + t_{operator}$$

The index for total real-time capability in the event of a fault (plcFCLAI) adds up to

$$plcFCLAI = plcDFAI + plcIFAI + plcSTSAI = 300 \text{ ms} + 12 \text{ ms} + 8 \text{ ms} + t_{operator} = 320 \text{ ms} + t_{operator}$$

By means of the metrics for real-time capability, different software implementations can be compared and the one with best real-time capability (lowest plcFCLAI-value) can be selected.

Since the analyzed program holds a recovery function for one fault only, the metrics for fault coverage are omitted in this example.

Metrics for the Effort to Increase Fault Coverage

The program is written in the programming language Structured Text. In order to measure the changes due to the addition of a recovery function, the metric Lines of Code (LOC) is applied. The total number of LOC of the overall program is 2511 without recovery function and in order to implement recovery, an additional 531 LOC are added. Thereby, the main function and the auxiliary functions are unchanged whereas every one of the modules has to be adapted.

The effort to implement recovery can be measured with the FHAI (cp. Formula (9)). In order to implement recovery for the fault "timing pusher", no inputs or outputs were added, no lines of the original code were adapted and no parts of the code were removed. Therefore, $FHAI_{new}$ and $FHAI_{old}$ (Formulas (11) and (13)) are applied in order to calculate the FHAI.

Since the auxiliary functions and the main function have not been changed at all, only the metric $FHAI_{initial}$ has to be calculated for those functions. The LOC of the main function and the nine auxiliary functions result in

$$FHAI_{initial\_main} = 214$$

$$FHAI_{initial\_auxiliary} = 34 + 106 + 33 + 39 + 48 + 292 + 47 + 22 + 49 = 670$$

Within the seven modules the LOCs of the original program were complemented by new LOCs, however, the original lines were not adapted. Thus, $FHAI_{new}$ and $FHAI_{initial}$ need to be calculated for the module functions.

$$FHAI_{new\_module} = 56 + 58 + 70 + 70 + 69 + 141 + 67 = 531$$

$$FHAI_{initial\_module} = 128 + 88 + 295 + 139 + 184 + 490 + 303 = 1627$$

Overall, $FHAI_{new}$ and $FHAI_{initial}$ add up to

$$FHAI_{new} = FHAI_{new\_module} = 531$$

$$FHAI_{initial} = FHAI_{initial\_main} + FHAI_{initial\_auxiliary} + FHAI_{initial\_module} = 214 + 670 + 1627 = 2511$$

Finally, the FHAI results in

$$FHAI = \frac{FHAI_{new\_module}}{FHAI_{new\_module} + FHAI_{initial}} = \frac{531}{2511 + 531} = \frac{531}{3042} = 0.17$$

This means that 17% of the software needed to be added in order to implement recovery from fault "timing pusher".

Compared to the first programs (PRG A.1 and PRG A.2) analyzed in Section 5.1 (cp. Figure 9), the real-time capabilities of the program introduced in this section (PRG B) are significantly lower. Programs PRG A.1 and PRG A.2 have a plcFCLAI value of 30 ms and 90 ms, whereas PRG B has a value of $(320 \text{ ms} + t_{operator})$. However, this is mainly due to safety requirements prohibiting an automatic start of the recovery $(t_{operator})$ and the chosen offset of 300 ms, which will be reduced in a future implementation. Since only one specific fault was considered in the OMAC program, no statement can be made about the metrics for fault coverage.

Results of the Evaluation

In this section the questions, whether the developed metrics are adequate to measure adaptivity in case of a fault and restart of an aPS after a shut down.

The evaluation of the four scenarios showed that the wide range of all four scenarios could be measured reaching from self-configuration based on hardware adaptation, i.e. added sensors (add-on and modification according to Gronau's classification), self-configuration on basis of virtual sensors to restartability after stop or emergency shut-down. The programming effort (fault handling adaptivity index) depends of course strongly on the numbers of faults being covered (cp. Table 2). Scenario 5.1 is the most sophisticated identifying the fault with redundant sensors, isolating it and automatically recovering by reversing the belt direction automatically to readjust the pusher and the work piece.

Table 2: Comparison of the metrics applied to the different application scenarios

	Metric		Case study/scenario			
			Section 5.1	Section 5.2	Section 5.3	Section 5.4
			PPU sorting	PPU crane	Tank filling level	xPPU sorting
	Adaptivity Concepts		FCA based fault detection and self-compensation	Restart after emergency shut down	Self-configuration with model based redundancy	Restart after fault from recovery state (OMAC SM)
	Real-time capabilities	plcDFAI	30 ms	-	t_{cycles}	300 ms
		plcIFAI	30 ms	-	0	12 ms
		plcSTSAI	30 ms	-	0	8 ms + $t_{operator}$
		plcFCLAI	90 ms	-	t_{cycles}	320 ms + $t_{operator}$
Applied metrics	Fault coverage	BECAI	0.5	-	-	-
		BENCAI	0.5	-	-	-
		FLCAI	0.167	-	0.33	-
	Minimal programming effort	MEICAI	0.33	-	-	-
		FHAI	0.61	-	0.31	0.17
	Minimal modelling effort	$V(G)_{old}$	2	4	-	-
		$V(G)_{new}$	5	6	-	-
		$FHAI_{model}$	0.857	0.4	-	-

"-": not applied.

Therefore, this scenario is the most challenging regarding the effort necessary to increase fault coverage, it includes restart (similar to 5.4). The other scenarios cover only limited aspects of the respective use case and therefore the effort is smaller, too. Scenario 5.3 only focuses on one fault and Scenario 5.2 and 5.4 only on the restart after an emergency shutdown or an inhibited work piece without identifying the reason of the fault.

Real time requirements are given by the technical process, which is controlled and have to be fulfilled. Same applies for metrics for real time capabilities to detect, isolate and compensate faults.

Conclusion/Summary/Outlook

Modularity has been identified as prerequisite for adaptive systems. This remains valid for Industry 4.0 or CPPS. The paper focuses on restart and self-configuration after a fault to increase OEE, i.e. to allow operation despite faulty devices and to detect and, isolate existing faults and adapt the machine or plant with self-configuration. Furthermore, the operator activities and interaction should be eased (cp 5.2 and 5.4). As a basis, eight use cases regarding fault handling from world market leading machine and plant manufacturing companies were introduced. Hierarchical fault handling is a frequent practice in special purpose machinery manufacturing as well as in plant manufacturing. The isolation of single faults and an automatic reaction to these faults needs to be implemented in order to be able to react to faults adaptively to reduce machine or plant shutdowns and additional manual interventions by operators in the future. In the domain of packaging plants for food & beverage, the OMAC state machine is already widely used which eases fault handling and restart (cp. 5.4 and case study E). Nevertheless, restart and self-configuration are limited due to the production process itself as well as redundant sensors or the availability of virtual sensors.

Selected metrics for adaptive automated Production Systems, selected metrics for real-time preservation in the event of a fault, fault coverage of a given automation system and the effort to extend the system, to increase fault coverage have been presented and evaluated with four scenarios ranging from no adaptation, adaptation add-ons, adaptation via modification to self-configuration. Future work will address on the one hand a more general challenge of software architecture of the effort to develop software for safety devices, e.g. protective covering as well as safety doors and on the other hand the development of more powerful metrics for adaptivity and Industry 4.0, which hopefully will allow to benchmark compliance of machine and plants to Industry 4.0.

ACKNOWLEDGEMENTS

We thank Professor Dr. H.C. Peter Göhner for his tireless commitment not just within FA 5.15 regarding the use of agents in automation and the successful realization of the demonstrator "MyJoghurt". We thank the FA 5.15 Agents in Automation Technology for the discussion on the topic of metrics in Industry 4.0 and CPPS as well as the companies for providing an insight into their challenges and software structures concerning architectures and fault handling as well as restart and recovery after faults. We also thank the DFG Priority Program 1593 "Design For Future—Managed Software Evolution" for fruitful discussion and for partially funding this research within the project MoDEMAS (grand number VO 937/20-1).

REFERENCES

1. Vogel-Heuser, B., Diedrich, C., Pantforder, D. and Gohner, P. (2014) Coupling Heterogeneous Production Systems by a Multi-Agent Based Cyber-Physical Production System. Proceedings of the 12th IEEE International Conference on Industrial Informatics (INDIN), Porto Alegre, 27-30 July 2014, 713-719.

2. VDI/VDE-Gesellschaft fur Messund Automatisierungstechnik (2015) Status Report Reference Architecture Model Industrie 4.0 (RAMI4.0). VDI e.V., Dusseldorf.

3. Vogel-Heuser, B., Fay, A., Schafer, I. and Tichy, M. (2015) Evolution of Software in Automated Production Systems—Challenges and Research Directions. Journal of Systems and Software (JSS), 110, 54-84. http://dx.doi.org/10.1016/j.jss.2015.08.026

4. Jazdi, N., Maga, C. and Gohner, P. (2011) Reusable Models in Industrial Automation: Experiences in Defining Appropriate Levels of Granularity. Proceedings of the 18th IFAC World Congress, Milano, 28 August-2 September 2011, 9145-9150.

5. Feldmann, S., Fuchs, J. and Vogel-Heuser, B. (2012) Modularity, Variant and Version Management in Plant Automation—Future Challenges and State of the Art. Proceedings of the 12th International Design Conference (DESIGN 2012), Dubrovnik, 25 May 2012, 1689-1698.

6. Feldmann, S., Legat, C. and Vogel-Heuser, B. (2015) Engineering Support in the Machine and Plant Manufacturing Domain through Interdisciplinary Product Lines: An Applicability Analysis. Proceedings of the 15th IFAC Symposium on Information Control in Manufacturing (INCOM), Ottawa, 11-13 May 2015, 211-218.

7. Katzke, U., Vogel-Heuser, B. and Fischer, K. (2004) Analysis and State of the Art of Modules in Industrial Automation. ATP International-Automation Technology in Practice International, 46, 23-31.

8. Vogel-Heuser, B., Fischer, J., Rosch, S., Feldmann, S. and Ulewicz, S. (2015) Challenges for Maintenance of PLC-Software and Its Related Hardware for Automated Production Systems—Selected Industrial Case Studies. Proceedings of the 31st IEEE International Conference on Software Maintenance and Evolution (ICSME), Bremen, 29 September-1 October 2015, 362-371.

9. Vyatkin, V. (2011) IEC 61499 as Enabler of Distributed and Intelligent

Automation: State-of-the-Art Review. IEEE Transactions on Industrial Informatics, 7, 768-781.

10. International Electronical Commission (2005) IEC International Standard IEC 61499-1: Function Blocks, Part 1: Architectures. IEC, Geneva.

11. Bonfè, M., Fantuzzi, C. and Secchi, C. (2013) Design Patterns for Model-Based Automation Software Design and Implementation. Control Engineering Practice, 21, 1608-1619. http://dx.doi.org/10.1016/j. conengprac.2012.03.017

12. Guttel, K., Weber, P. and Fay, A. (2008). Automatic Generation of PLC Code beyond the Nominal Sequence. IEEE International Conference on Emerging Technologies and Factory Automation (ETFA), Hamburg, 15-18 September 2008, 1277-1284.

13. Barbieri, G., Battilani, N., and Fantuzzi, C. (2015). A PackML-Based Design Pattern for Modular PLC Code. IFAC-Papers on Line, 48, 178-183.http://dx.doi.org/10.1016/j.ifacol.2015.08.128

14. Weihenstephan Standards. http://www.weihenstephaner-standards.de

15. OMAC. http://www.omac.org/content/packml

16. Vogel-Heuser, B., Legat, C., Folmer, J. and Rosch, S. (2014) Challenges of Parallel Evolution in Production Automation Focusing on Requirements Specification and Fault Handling. Automatisierungstechnik (at), 755-826. http://dx.doi.org/10.1515/auto-2014-1111

17. Schutz, D., Wannagat, A., Legat, C. and Vogel-Heuser, B. (2013) Development of PLC-Based Software for Increasing the Dependability of Production Automation Systems. IEEE Transactions on Industrial Informatics, 9, 2397-2406.

18. Bergagard, P., Falkman, P. and Fabian, M. (2015) Modeling and Automatic Calculation of Restart States For an Industrial Windscreen Mounting Station. Proceedings of the IFAC Symposium on Information Control Problems in Manufacturing (INCOM), Ottawa, 11-13 May 2015, 1030-1036.

19. Vogel-Heuser, B., Schutz, D., Frank, T. and Legat, C. (2014) Model-Driven Engineering of Manufacturing Automation Software Projects—A SysML-Based Approach. Mechatronics, 24, 883-897. http://dx.doi. org/10.1016/j.mechatronics.2014.05.003

20. Priego, R., Schutz, D., Vogel-Heuser, B. and Marcos, M. (2015) Reconfiguration Architecture for Updates of Automation Systems during Operation. Proceedings of the 20th IEEE International Conference on Emerging Technologies and Factory Automation (ETFA), Luxembourg,

8-11 September 2015, 1-8.

21. Andersson, K., Lennartson, B., Falkman, P. and Fabian, M. (2011) Generation of Restart States for Manufacturing Cell Controllers. Control Engineering Practice, 19, 1014-1022. http://dx.doi.org/10.1016/j. conengprac.2011.05.013

22. Andersson, K., Lennartson, B. and Fabian, M. (2010) Restarting Manufacturing Systems; Restart States and Restartability. IEEE Transactions on Automation Science and Engineering, 7, 486-499. http://dx.doi.org/10.1109/TASE.2009.2034136

23. Bergagard, P. (2015) On Restart of Automated Manufacturing Systems, Ph.D. Dissertation, Department of Signals and Systems, Chalmers University of Technology, Goteborg.

24. Andersson, K., Richardsson, J., Lennartson, B. and Fabian, M. (2009) Coordination of Operations by Relation Extraction for Manufacturing Cell Controllers. IEEE Transactions on Control Systems Technology, 18, 414-429.

25. Bergagard, P. and Fabian, M. (2013) Calculating Restart States for Systems Modeled by Operations Using Supervisory Control Theory. Machines, 1, 116-141. http://dx.doi.org/10.3390/machines1030116

26. Lennartson, B., et al. (2010) Sequence Planning for Integrated Product, Process and Automation Design. IEEE Transactions on Automation Science and Engineering, 7, 791-802. http://dx.doi.org/10.1109/TASE.2010.2051664

27. Bengtsson, K., et al. (2012) Sequence Planning Using Multiple and Coordinated Sequences of Operations. IEEE Transactions on Automation Science and Engineering, 9, 308-319. http://dx.doi.org/10.1109/ TASE.2011.2178068

28. Ladiges, J., Fay, A., Haubeck, C. and Lamersdorf, W. (2013) Operationalized Definitions of Non-Functional Requirements on Automated Production Facilities to Measure Evolution Effects with an Automation System. IEEE International Conference on Emerging Technologies and Factory Automation (ETFA), Cagliari, 10-13 September 2013, 1-6.

29. Gronau, N., Lammer, A. and Andresen, K. (2007) Entwicklung wandlungsfahiger Auftragsabwicklungs systeme. GITO-Verlag, Berlin.

30. Wiendahl, H.-P., et al. (2007) Changeable Manufacturing—Classification, Design and Operation. CIRP Annals Manufacturing Technology, 56, 783-809. http://dx.doi.org/10.1016/j.cirp.2007.10.003

31. Raibulet, C. and Masciadri, L. (2009) Towards Evaluation Mechanisms

for Runtime Adaptivity: From Case Studies to Metrics. Computation World: Future Computing, Service Computation Cognitive, Adaptive, Content, Patterns, Athens, 15-20 November 2009, 146-152. http://dx.doi.org/10.1109/computationworld.2009.89

32. Fischer, J., Friedrich, D. and Vogel-Heuser, B. (2015) Configuration of PLC Software for Automated Warehouses Based on Reusable Components—An Industrial Case Study. Proceedings of the 20th IEEE International Conference on Emerging Technologies and Factory Automation (ETFA), Luxembourg, 8-11 September 2015, 1-7.

33. German Institute for Standardization—DIN Deutsches Institut fur Normung e. V. (1985) DIN 25419: Event Tree Analysis; Method, Graphical Symbols and Evaluation. DIN, Berlin.

34. Friedrich, A. and Gohner, P. (2015) Fault Diagnosis of Automated Systems Using Mobile Devices. Proceedings of the 20th IEEE International Conference on Emerging Technologies and Factory Automation (ETFA), Luxembourg, 8-11 September 2015, 1-8.

35. Fuchs, J., Feldmann, S., Legat, C. and Vogel-Heuser, B. (2014) Identification of Design Patterns for IEC 61131-3 in Machine and Plant Manufacturing. Proceedings of the 19th IFAC World Congress (IFAC 2014), Cape Town, 24-29 August 2014, 6092-6097.

36. Vogel-Heuser, B. (2014) Usability Experiments to Evaluate UML/SysML-Based Model Driven Software Engineering Notations for Logic Control in Manufacturing Automation. Journal of Software Engineering and Applications, 7, 943-973. http://dx.doi.org/10.4236/jsea.2014.711084

37. Frey, G., Litz, L. and Klockner, F. (2000) Complexity Metrics for Petri Net Based Logic Control Algorithms. IEEE International Conference on Systems, Man, and Cybernetics, Nashville, TN, 8-11 October 2000, 1204-1209. http://dx.doi.org/10.1109/icsmc.2000.886016

38. Chidamber, S.R. and Kemerer, C.F. (1994) A Metrics Suite for Object Oriented Design. IEEE Transactions on Software Engineering, 20, 476-493. http://dx.doi.org/10.1109/32.295895

39. Michura, J. and Capretz, M.A.M. (2005) Metrics Suite for Class Complexity. IEEE International Conference on Information Technology: Coding and Computing (ITCC), Las Vegas, NV, 4-6 April 2005, 404-409. http://dx.doi.org/10.1109/itcc.2005.193

40. McCabe, T.J. (1976) A Complexity Measure. IEEE Transactions on Software Engineering, SE-2, 308-320. http://dx.doi.org/10.1109/TSE.1976.233837

41. Witsch, D. and Vogel-Heuser, B. (2011) PLC-Statecharts: An Approach to Integrate UML-Statecharts in Open-Loop Control Engineering—Aspects on Behavioral Semantics and Model-Checking. Proceedings of the 18th IFAC World Congress, Milano, 28 August-2 September 2011, 7866-7872.

42. German Institute for Standardization—DIN Deutsches Institut fur Normung e. V. (2009) DIN EN 13128: Safety of Machine Tools—Milling Machines (Including Boring Machines). DIN, Berlin.

Chapter 4

A REFERENCE-PULSE GENERATOR FOR MOTION CONTROL SYSTEM

Nguyen Hoang Giap[1], Jin-Ho Shin[2], Won-Ho Kim[2]

[1]Department of Intelligent System Engineering, Graduate School of Dong-eui University, Busan, Korea

[2]Department of Mechatronics Engineering, Dong-eui University, Busan, Korea

ABSTRACT

This paper introduces a reference-pulse interpolator motion control system, which can be applied for computer numerical control (CNC) machine tools. The interpolation is calculated in DSP and the independent pulse generator modules are performed in FPGA, which can generate precise reference velocity profile and eliminate the path error of reference pulse interpolation. The proposed methodology has many advantages over existing reference-pulse interpolator controller, such as: real-time, extensibility, high flexibility and high precision motion profile planning.

INTRODUCTION

In the last few decades, motion control has developed steadily with the contribution of recent advanced control theories and technologies [1] . Along with the intensive applications of motion control, many researches have been conducted to enhance the performance of the motion control system [2] - [4] . The development of Field- Programmable Gate Array (FPGA) makes it easy to implement motion control applications due to programmable hard-wired feature, parallel processing architecture, short design cycle, low power consumption and high density [5] . Therefore FPGA is conventionally used to develop compact, low-price, high volume, and high performance motion control system [6] . However, FPGA-based controller with only hard-wired logic is not flexible and expandable for designing complicated algorithms and structures of motion control system, such as user's defined motion profile, motion blending, advanced closed-loop control algorithms. Therefore, a new motion control hardware architecture was proposed by X. Shao, D. Shun, and J. K. Mills [7] to improve motion performance of high speed robotic manipulators. In this research, a DSP is used for dynamic compensation, trajectory planning, and the control loop is implemented in the FPGA. In [8] , a System-on-a Programmable-Chip (SoPC) with an embedded processor is implemented in FPGA to carry out motion profile generation and position control algorithm, while current vector control is performed in FPGA. This technology enables software and hardware co-design to be parallel processing, hence increases the system performance and flexibility [8] - [10] . However, all of above combinations are just developed in closed-loop control types.

Open-loop motion controller type generates reference signals, which is generally used for CNC systems such as water-jet cutting, machine tools, laser-beam cutters and welders. The reference signals from the controller can be transmitted as a sequence of reference pulse or as binary word in a sampled-data system [11] . The interpolator and signal generators are normally implemented in PC-based system with assembly language in order to save memory data and improve computing speed [11] [12] . In this system, the digital differential analyzer (DDA) method [11] [13] or stairs approximation method [12] [14] is used in hardware interpolators. Although the program design of geometric path planning is simple in PC-based control, the restriction of interpolation execution time of PC-based system makes it difficult to expand multi-axis motion system. Moreover, the design of interpolator and pulse generation based on one interrupt clock employed in these methods causes path error and cannot generate constant instantaneous velocity [12] . In [12] , a simple motion control approach is proposed by using two separate interrupt interval times at each interpolation step to generate constant velocity along geometric path. Although the precision is improved, this method still has error of half basic length unit.

In this paper we propose a new structure to overcome those drawbacks. Firstly, DSP is used to employ complicated calculation of the geometric path planning and motion profile generation. The order of velocity profile can be expanded to reduce jerk and vibration residue. Secondly, the binary velocity data of individual axis obtained from DSP are sent simultaneously to FPGA through Universal Host Port Interface (UHPI) at each sampling clock. A Master-Slave Wisbone interface (MSWI) module is designed in FPGA to communicate with DSP and pulse generator module through a First-In-First-Out (FIFO) memory buffer controller. Pulse generator modules with high speed clock designed in FPGA receive the velocity data from DSP and generate output pulse independently for each axis. The combination of interpolation calculation in DSP and independent pulse generator module in FPGA, along with the synchronous communication hardware structure, makes it possible to design a synchronous motion control system and eliminate the path error of reference pulse interpolation.

DESIGN OF SYNCHRONOUS MOTION CONTROL SYSTEM

Architecture of Synchronous Motion Control System

The reference pulse motion controller generates a sequence of pulses representing position and velocity to be controlled, which is usually based on an iterative technique controlled by an interrupt clock.

In order to overcome the drawbacks of this kind of motion control, we proposed a new synchronous interpolation hardware and software structure, where the motion of each axis is independent with the system clock rate as shown in Figure 1. Consequently, the constant instantaneous velocity along the geometric path is generated and the error of pulse generator is eliminated.

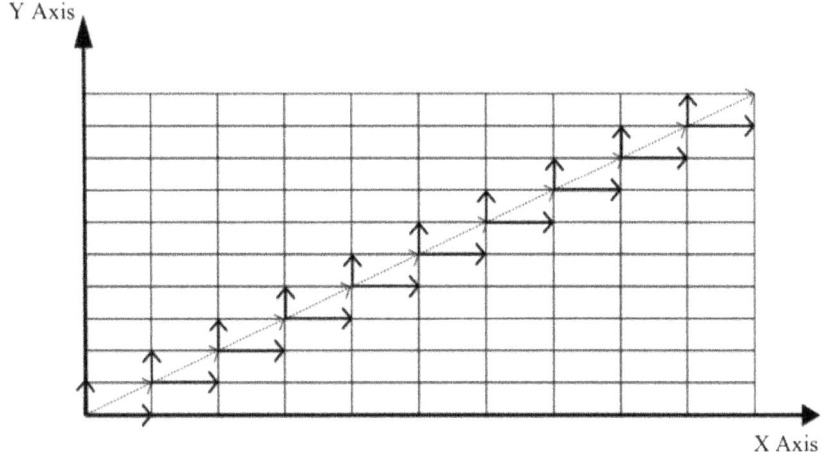

Figure 1: Proposed interpolation scheme.

The proposed controller is developed based on a high performance DSP and an FPGA. The geometric path planning and motion profile generation are calculated by DSP. At each interrupt clock, the interpolated velocity data of each coordinate are transferred to the pulse generator module through the MSWI. The MSWI module, designed in FPGA includes a Wishbone bus arbitration, a Wishbone master interface, a Wisbone slave interface for internal SRAM memory control, and Wishbone slave interfaces for connecting with the pulse generator module s through a FIFO buffer memory controller.

The FIFO buffer memory controller includes a timer and a shift register control. The signal i_clk and r_clk_full represent the input clock and output for timer, respectively. The frequency of r_clk_full signal is equivalent to the clock interrupt of DSP. This timer ensures the velocity data to be transferred between DSP and pulse generator module of FPGA synchronously. Memory buffer contains velocity data r_data to gradually transfer to the pulse generator module at every sampling clock. Then these velocity data are accumulated in the pulse generator module to generate reference pulse. A shift register in the stack occurs when there exists velocity input data i_wr_data from the host controller and i_time_trigger signal is active. The i_time_trigger signal indicates that the buffer memory is not empty and timer signal i_clk is active.

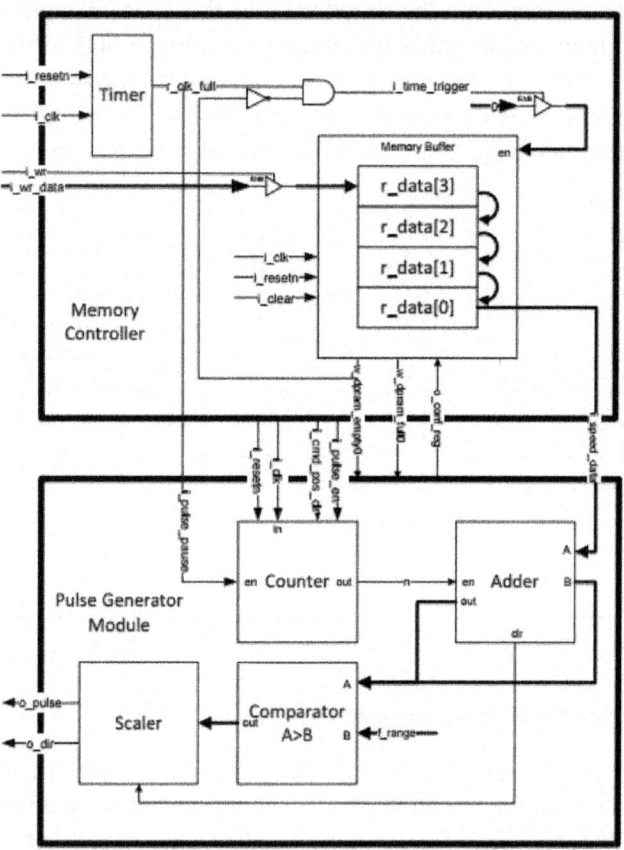

Figure 2: Pulse generator module.

The bus arbitration in MSWI is used to inter-connect the bus control signals between master and slave module. This structure enables the interpolator velocity data to be transferred from DSP to pulse generator module s simultaneously, and the corresponding pulse outputs generated for all coordinates are synchronous at each interval clock.

Pulse generator module is connected with the Wishbone slave interface through a FIFO buffer memory controller as shown in Figure 2. This design protects the binary velocity data from the host controller to be overflowed. Consequently, the proposed controller is capable of generating synchronous and precise reference pulse signal for motion control system.

Pulse Generator Module

Pulse generator module generates a sequence of signals o_pulse and o_dir representing reference position and moving direction of motion system, respectively. At each sampling time, the binary velocity data i_speed_data are stored in memory buffer for pulse generation process.

Since the pulse signal is equivalent to one basic length unit (BLU), the pulse frequency is proportional to velocity, and the number of pulse is equivalent to the integration of velocity over time unit as in Equation (1):

$$P = \int_0^t BVU \times V dt \qquad (1)$$

$$BVU = \frac{f_{CLK}}{f_{RANGE}} \qquad (2)$$

where P denotes the number of reference pulse output, BVU represents the basic velocity unit which is proportional to the frequency f_{clk} of timer signal i_clk, and V is the reference velocity (pulse/sec), f_{RANGE} is the frequency unit defines the input factor for Comparator block.

Equation (1) can be rewritten in discrete type as:

$$P = \frac{\sum_{n=0}^{N} nV}{f_{RANGE}} \qquad (3)$$

$$n = f_{CLK} t \qquad (4)$$

where n is the output of Counter block.

The Counter and Adder block in pulse generator module acts as the velocity integrator.

The integrator value $\sum_{n=0}^{N} nV$ is compared with f_{RANGE} factor in the Comparator block. An output pulse pulse is toggle when $\sum_{n=0}^{N} nV \geq f_{RANGE}/2$.

GEOMETRIC PATH PLANNING AND MOTION PROFILE GENERATION

Motion planning, including geometric path planning G(p) and motion profile generation p(t), plays an important role in machine control. With the rapid development of microprocessor technology, motion planning becomes flexible to employ [10] . In this section, we will discuss about the effective method to design a DSP-based motion planning module that can be applied for reference pulse generator.

Motion Profile Generation

Motion profile is expressed as a parametric function of time p(t), which provides the corresponding desired position and velocity at each instant. S-curve trajectory or seven-segment trajectory is used in motion control system to provide a smooth motion profile by adopting a continuous, linear piece-wise acceleration profile. In this manner, the resulting velocity is composed by linear segment connected by parabolic blends. Since the jerk is characterized by a step profile, the stress load by this motion profile is reduced.

The seven main parts of S-curve profile can be determined from the input coordinate motion data, as follows:

1st, 3rd, 5th, and 7th segments:

$$p(t) = p_0 + \dot{p}_0 t + \frac{1}{2} \ddot{p}_0 t^2 + \frac{1}{6} j_{max} t^3 \qquad (5)$$

$$\dot{p}(t) = \dot{p}_0 + \ddot{p}_0 t + \frac{1}{2} j_{max} t^2 \qquad (6)$$

2nd, and 6th segments:

$$p(t) = p_0 + \dot{p}_0 t + \frac{1}{2} \ddot{p}_{max} t^2 \qquad (7)$$

$$\dot{p}(t) = \dot{p}_0 + \ddot{p}_{max} t \qquad (8)$$

4th segments:

$$p(t) = p_0 + \dot{p}_{max} t \qquad (9)$$

$$\dot{p}(t) = \dot{p}_{max} \qquad (10)$$

where $p(t)$ denotes the instant position, p_0 is initial position, $\dot{p}_{max}$ is maximum velocity, and $\ddot{p}_{max}$ is maximum acceleration of geometric path, respectively.

Geometric Path Planning

Generally, geometric path planning including linear interpolation is common applied in industrial manufacture. Geometric path planning generates the reference position $G(p)$ and velocity command $\dot{G}(p)$ for individual coordinate which is used for reference pulse generator. The geometric path can be considered as a parametric vector function which represents the trajectory when the motion profile $p(t)$ moves over some interval of coordination as following:

$$G(p) = \begin{bmatrix} G_X(p) & G_Y(p) & G_Z(p) \end{bmatrix} \tag{11}$$

Linear interpolant is a straight line between two given points (G_{X0}, G_{Y0}) and (G_{X1}, G_{Y1}). For a value G_X in the interval (G_{X0}, G_{X1}), the value G_Y along the straight line is calculated from the following equation:

$$\frac{G_Y - G_{Y0}}{G_X - G_{X0}} = \frac{G_{Y1} - G_{Y0}}{G_{X1} - G_{X0}} \tag{12}$$

The design of linear interpolation function in microprocessor is straightforward. The displacement of linear trajectory can be calculated as following:

$$p_l = \sqrt{(G_{X1} - G_{X0})^2 + (G_{Y1} - G_{Y0})^2} \tag{13}$$

The interpolated position and velocity of reference path can be presented as following:

$$G_X = G_{X0} + p\cos\theta_l \tag{14}$$

$$\dot{G}_X = \dot{G}_{X0} + \dot{p}\cos\theta_l \tag{15}$$

$$G_Y = G_{Y0} + p\sin\theta_l \tag{16}$$

$$\dot{G}_Y = \dot{G}_{Y0} + \dot{p}\sin\theta_l \tag{17}$$

It can be seen from the above calculations, the instantaneous position and velocity data of individual axis can be implemented comprehensively in DSP by some calculations of motion profile generation and geometric path planning process.

EXPERIMENTAL RESULTS

The overall motion control system shown in Figure 3 consists of an embedded motion control board, a host PC to communicate with the board, 4 servo drivers to drive 4 AC motors of gantry mechanical system. In this experiment, only a vertical axis and a horizontal axis are used for linear move. The developed embedded motion control board, showed in Figure 4, includes a high performance DSP TMS320F2812 operating at 150 MHz and an FPGA XC3S1000 that has 1M system gates, 17,280 logic cells, and 391 I/O pins. The summary of the device utilization characteristic for the circuit of the FPGA is given in Table 1. The line driver 75ALS174A is used to convert the reference single-end pulse/dir signal from the FPGA to differential-end signal. The servo driver is set to operate in position mode. A quadrature encoder module (QEM) is designed in FPGA to observe the actual position of mechanical system. The line receiver is used to convert differential-end signal from rotary encoder of motor to single-end signal for QEM. The velocity profile data is send periodically from DSP to FPGA at the sampling clock frequency of 2 KHz. The pulse generator module in FPGA is able to generate pulse at maximum speed of 4 Mpps. The motion control board implemented in this paper can control system with up to 4 axes.

Figure 5 shows the reference pulses of proposed motion control board, which represents the linear interpolation from (0.00 mm, 0.00 mm) to (500.00 mm, 800.00 mm) for the required vector velocity of 400.00 mm/s. The

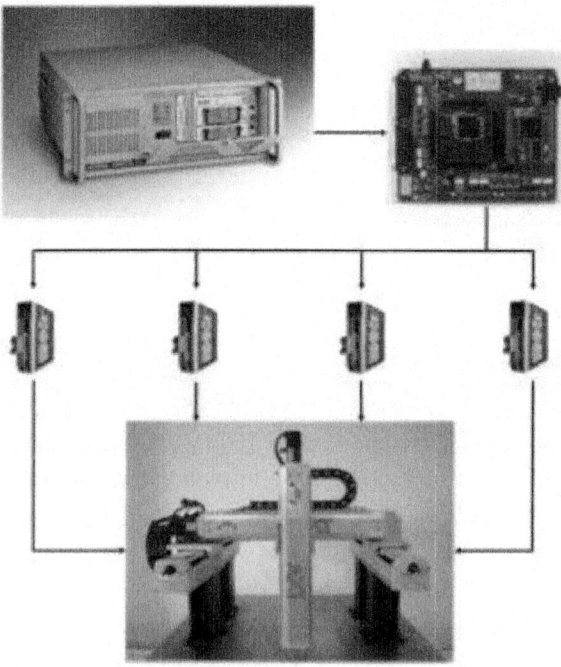

Figure 3: Multiple axis motion control system.

Figure 4: Developed motion control board.

Table 1: Device utilization summary

Device Utilization Summary	Logic Utilization		
	Total	Used	%
Total Number Slice Registers	15,360	3594	23
Number of 4 input LUTs	15,360	5738	37
Number of occupied Slices	7680	3518	45
Total Number of 4 input LUTs	15,360	5975	38
Number of bonded IOBs	333	219	65
Number of BUFGMUXs	8	3	37

BLU of pulse generator module is 0.0001 mm. Figures 6-8, present the velocity profile, position profile and geometric path planning generated by linear interpolation. The actual position and velocity of each axis are gathered by QEM module from servo drivers to compare with the result of proposed reference pulse generators.

Meanwhile, due to the combination of interpolation calculation from DSP and independent pulse generator module from FPGA, the instantaneous vector velocity of linear path generated by proposed method can keep a constant value, and the reference path error is 0. The experimental results in Figure 5 shows that the pulses generated from two axes are uniform and synchronous.

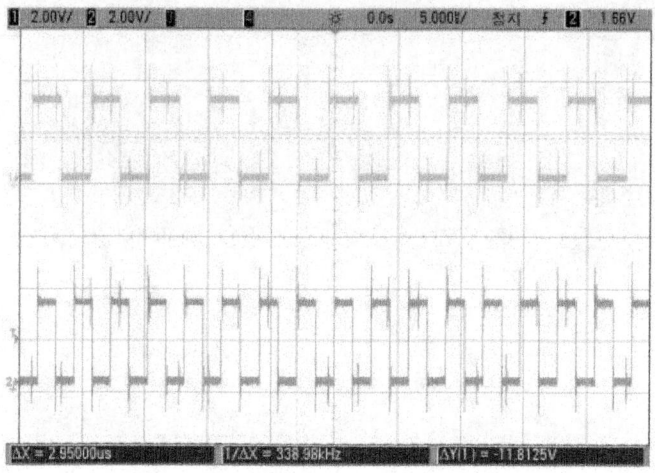

Figure 5: Reference pulse of X and Y axis along linear path using proposed controller.

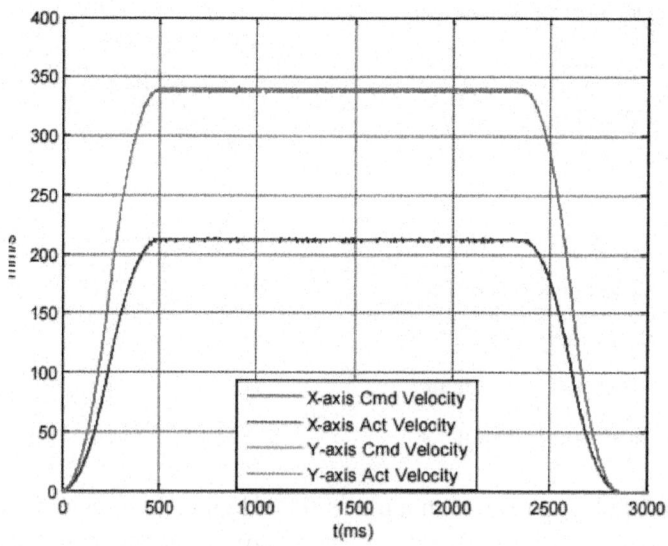

Figure 6: Single axis velocity of linear interpolation.

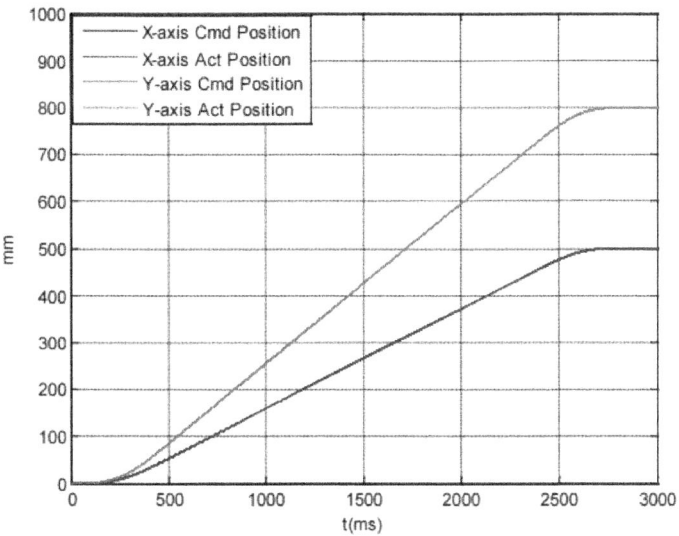

Figure 7: Single axis position of linear interpolation.

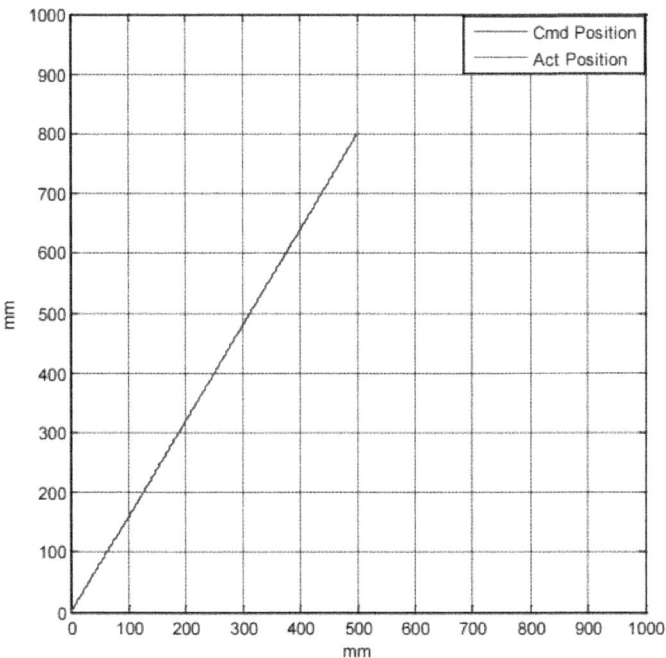

Figure 8; Position of linear interpolation.

The velocity profiles in Figure 6 and the position tracking in Figure 7 validate the calculation of motion profile generation in the DSP.

The experimental results prove the synchronous performance of developed controller applied in an industrial motion system. The flexibility and the open design of the motion board make it easy to develop and configure advanced motion control algorithm.

CONCLUSION

In this paper, we introduced a new synchronous reference pulse interpolator methodology for motion control system. The combination of proposed interpolation calculation in DSP and independent pulse generator module in FPGA can generate precise reference velocity profile and eliminate the path error of reference pulse interpolation. Therefore the proposed motion controller is applicable in high precision motion control system like CNC machine, planning machine, robot control.

ACKNOWLEDGEMENTS

This work was supported by Dong-Eui University Foundation Grant (grant number 2014AA492).

REFERENCES

1.   Desborough, L. and Miller, R. (2002) Increasing Customer Value of Industrial Control Performance Monitoring— Honeywell Experience. In: Rawlings, J.B., Ogunnaike, B.A. and Eaton, J.W., Eds., 6th International Conference Chemical Process Control, AIChE Symp., Series 326, New York, AIChE.

2.   Volpe, R. (1993) Task Space Velocity Blending for Real-Time Trajectory Generation. Proceedings of IEEE International Conference on Robotics and Automation, Atlanta, GA, 2-6 May 1993, 680-687. http://dx.doi.org/10.1109/ROBOT.1993.291880

3.   Macfarlane, S. and Croft, E.A. (2003) Jerk-Bounded Manipulator Trajectory Planning: Design for Real-Time Applications. IEEE Transactions on Robotics and Automation, 19, 42-52. http://dx.doi.org/10.1109/TRA.2002.807548

4.   Meckl, P.H., Arestides, P.B. and Woods, M.C. (1998) Optimized S-Curve Motion Profiles for Minimum Residual Vibration. Proceedings of the 1998 American Control Conference, Philadelphia, PA, 21-26 June 1998, 2627-2631.

5. Kung, Y.-S., Tseng, K.-H. and Tai, T.-Y. (2006) FPGA-Based Servo Control IC for X-Y Table. Proceedings of IEEE International Conference on Industrial Technology, Mumbai, 15-17 December 2006, 2913-2918.

6. Cho, J.U., Ngoc, Q.L. and Jeon, J.W. (2009) An FPGA-Based Multiple-Axis Motion Control Chip. IEEE Transactions on Industrial Electronics, 56, 856-870.http://dx.doi.org/10.1109/TIE.2008.2004671

7. Shao, X., Shun, D. and Mills, J.K. (2006) A New Motion Control Hardware Architecture with FPGA-Based IC Design for Robotic Manipulators. Proceedings of IEEE International Conference on Robotics and Automation, Orlando, FL, 15-19 May 2006, 3520-3525.

8. Xu, J., You, B. and Ma, L. (2008) Research and Development of DSP Based Servo Motion Controller. Proceedings of the 7th World Congress on Intelligent Control and Automation, Chongqing, 25-27 June 2008, 7720-7725.

9. Naouar, M.W., Monmasson, E., Naassani, A.A., Belkhodja, I.S. and Patin, N. (2007) FPGA-Based Current Controllers for AC Machine Drives—A Review. IEEE Transactions on Industrial Electronics, 54, 1907-1925. http://dx.doi.org/10.1109/TIE.2007.898302

10. You, B., Li, D. and Liu, S. (2007) Design of DSP-Based Open Control System for Industrial Robot. Proceedings of the IEEE International Conference on Automation and Logistics, Jinan, 18-21 August 2007, 1585-1590.

11. Koren, Y. and Masory, O. (1981) Reference-Pulse Circular Interpolators For CNC Systems. Journal of Manufacturing Science and Engineering, 103, 131-136.http://dx.doi.org/10.1115/1.3184454

12. Lho, T.J. and Kim, J.Y. (2000) A Study on Development of a PC-Based Software Interpolator for Two-Axis CNC Systems. TIT Research Journal, 3, 103-121.

13. Koren, Y. (1976) Interpolator for a Computer Numerical Control System. IEEE Transactions on Computers, C-25, 32-37. http://dx.doi.org/10.1109/TC.1976.5009202

14. Lee, B.J. and Nho, T.S. (1982) Linear and Circular Interpolation for 2-Dimension Contouring Control. Journal of Korean Society of Mechanical Engineers, 341-345.

Chapter 5

DESIGN AND IMPLEMENTATION OF FFPIV SCHEME FOR CLOSED LOOP MOTION CONTROLLER

Ngo Ha Quang Thinh[1], Jae-Gark Choi[2], Won-Ho Kim[3]

[1]Department of Intelligent System Engineering, Graduate School of Dong-eui University, Busan, Korea

[2]Department of Computer Engineering, Dong-eui University, Busan, Korea

[3]Department of Mechatronics Engineering, Dong-eui University, Busan, Korea

ABSTRACT

To satisfy the requirements of motion control for industrial machine, a multi-axis motion controller based on DSP is developed in this paper. The motion controller consists of DSP which plays a main role in this design; DPRAM to make sure the rapid and reliable communication with host; FPGA to handle the task of address decoder and receiving feed-back encoder signal; and several peripheral logic circuits. In the part of hardware design, overall structure of motion control system is presented. Then, the Feed-Forward Proportional-Integral-Velocity (FFPIV) scheme which introduces K_V in term of velocity loop to achieve the accurate, smooth and real-time response is proposed in the software developing part. The experiment data are carried out to indicate that this motion controller has advantages of superior performance and highly machining precision.

INTRODUCTION

For many years, motion controllers have been extensively applied in industrial automation systems. A motion controller is a kernel control device in electronic and mechanical manufacturing. Classifying on communication interfaces between hosts and controllers, motion controllers fall into sub-categories such as parallel communication like PCI-bus and ISA-bus, serial ones like RS232 and Ethernet protocol. Although serial-bus-based systems have lower prices and flexibility due to popular Ethernet standard [1] [2] , these systems are not guaranteed stability and/or desired performance, especially high traffic on the bus. Conversely, parallel-bus-based applications present good performance in term of high speed and high accuracy. However, ISA-bus-based control systems meet great drawbacks: tightly bounded to host architecture which is normally old design, lack of geographical addressing to add new devices and low response speed. Hence, many works have developed the PCI-bus-based motion control systems using specific chips or general purpose microprocessors. In [3] , authors investigated a simple motion controller based on LM628 microprocessor. This controller can simplify the hardware circuit and reduce the developing costs. On the other hand, it is only suitable for medium-sizedand mini-type control systems for the few axes. Furthermore, more axis controlling can be achieved by using DSP chip as the main element of motion controller.

Various researchers focus on developing DSP-based motion control systems. Authors in [4] [5] proposed a multi-axis servo motor motion controller based on DSP. In this design, DSP chip communicates with upper computer through Dual Port RAM and receive feedback signal from servo driver. Then, this controller calculates the error between theoretical interpolation position and actual position, and operates PID algorithm to achieve target position. Besides, Komin and Tanta-ngai in [6] also presented DSP-Based motion controller system that is able to control up to eight motors at the same time. Motion controller collects G-code from PC, computes path profile and interprets to command in order to control motor drive. By using profile calculation method, the developed system provides smooth movement for various path types. Chunhao et al. in [7] introduced a simple but effective motion controller based on the industrial PC. The advantages of this system are the low price and welladapted design to integrate into packaging, printing, textile and assembling industry.

Recently, several topics of advanced control are applied in motion controller such as a modified fuzzy logic control in automotive system [8] , fuzzy PID with feed-forward control strategy in CNC machine [9] [10] , fuzzy PI/PD-based control scheme in tracking applications including disturbance rejection and external loading [11] [12] , self-tuning fuzzy PID through continuous updating approach of output scaling factor [13] , a hybrid fuzzy bang-bang controller to improve the motor behavior [14] [15] or fuzzy PID with a class of gain matrices depending on the manipulator states [16] .

The other strategic approach to enhance the performance of manufacturing machine is how to lessen the tracking error. In modern and practical environment of production, existing frictions degrade the manufacturing precision. Researchers in [17] study two friction models in machine tools namely; static friction and Generalized Maxwell-slip model. Both identified friction models can be applied in feed-forward compensation scheme to improve tracking and positioning performance of machine tools. For velocity-controlled motion system, velocity-based friction compensation structure with additional velocity input in [18] is developed to overcome the adverse effects induced by friction to an extent. The friction compensation consists of the given position commands, velocity commands and an integrated friction model. The results prove that three-axis machine can reduce significantly the absolute contouring errors during motion. Besides, the uncertainties of parameter variation and measurement problems lead to decrease the motion accuracy. In [19], a state-space disturbance observer is successfully applied to estimate and compensate for these uncertainties. Experimental results indicate that the roundness error has been considerably reduced. For the other implementation issue, authors in [20] gave out a frequency domain based method for controller design in nonlinear system. This approach optimally designs a feed-forward friction compensator. It is suggested that in frequency domain the design yields a tool to fast and easily control the friction in practice.

In this paper, a novel DSP-based motion controller for industrial machine that implements feed-forward PIV algorithm is proposed. This controller can significantly lessen the tracking error that is the main factor of high precision machining, and reject the unknown disturbance of model. Therefore, the motion controller which consists of design of DSP and FPGA can be used in manufacturing broadly.

STRUCTURE OF MOTION CONTROL SYSTEM

The overall structure of motion control system is illustrated in Figure 1. The host IPC is Industrial PC with windows operating system. Through PCI bus, more than one industrial motion controller can communicate with host. Moreover, a motion controller can control up to 8 servo drivers simultaneously. Hence, the task of industrial motion controller is to perform arithmetic calculation operations with data from host and receive the displacement from servo driver. Additionally, this controller distributes the control value to servo motor in order to form the closed-loop control.

HARDWARE DESIGN OF MOTION CONTROLLER

In the motion controller, main hardware components are displayed in Figure 2. The core of motion controller is a Digital-Signal-Processor, which is explored to use the powerful computation function.

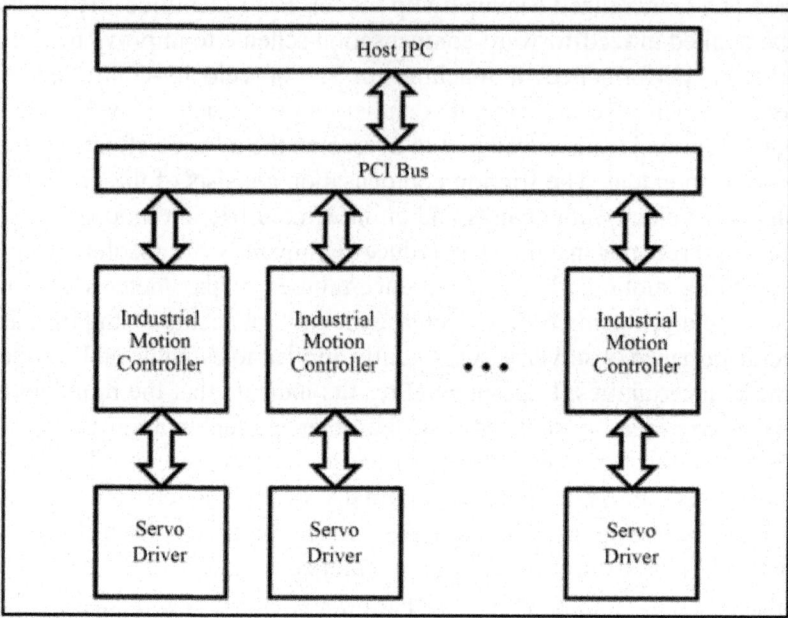

Figure 1: Overall structure of motion control system.

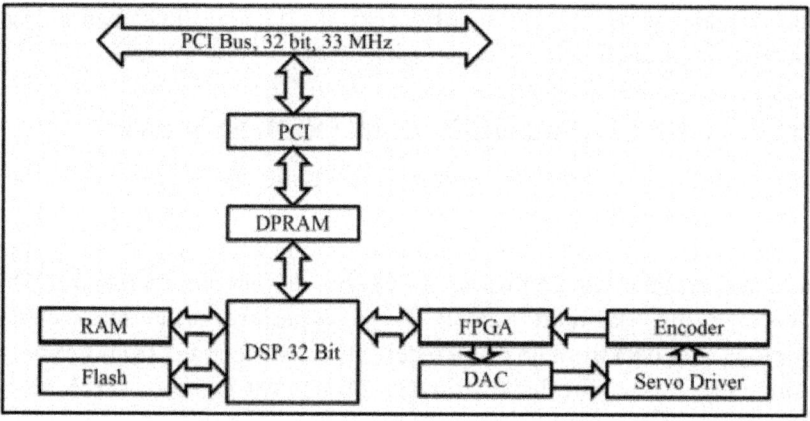

Figure 2: Hardware design of proposed motion controller.

The DSP completes most of tasks such as real-time communication with PC, motion profile generator, velocity and position control of servo motor, monitoring condition of servo motor and external signal detection. In addition, this motion controller also includes Digital-Analog-Converter, Dual-Port-RAM, Synchronous-DRAM, Flash, FPGA, PCI-interface-chip and other peripheral circuits. Whilst DPRAM not only holds the communication interfaces between controller and host but also supports sufficient buffer, FPGA is responsible as address decoder. Furthermore, PCI-interface-chip is a 32-bit PCI-bus target interface chip to ensure the PCI industry standard. In order to support memory spaces, RAM and Flash are designed to store data (position, velocity, sensor signal, parameters of controller, etc.).

When the controller works, host transmits user-specified data from DPRAM to DSP to provide the initial information. In a cycle of servo, DSP achieves the feedback signals, for instance, position and speed from servo motor. Therefore, it computes the differences between hypothetical values and real values, and executes the control strategies to acquire the necessary adjustments. Then, the output results are carried on digital analogy conversion by DAC chip and amplification by amplifier circuit. Finally, the output signal will be sent to the servo driver in order to control motor.

DSP Module

In recent years, DSP is used in digital control system widely, especially in motion control system. The reasons to choose DSP as the central component are primarily in processing ability, on-chip memory, interface and other on-chip resources. It is a high performance 32-bit floating-point digital signal processor, which is appropriate for industrial control application. External working clock's frequency can be up to 300 MHz which allows the execution of eight instructions in parallel each cycle.

FPGA Module

The design of FPGA focuses on input and output module to handle the communication in motion controller: DSP chip, D/A conversion, encoder signal and other feedback signals. It receives the data from servo driver; count a number of pulses and feedback to DSP. The interrupt is generated whenever positive/negative limit signals, home signal and alarm signal appear to be processed. The operator to combine among these signals is OR. When the rising edge of signal comes to FPGA chip, there will be a trigger signal to DSP in order to generate the interrupt service routine.

Communication Modules

The computer in early period used ISA-bus to control the motor. Nonetheless, the response speed was slow, and real-time performance was not guaranteed in those days. Since PCI-based control system was developed, motion control system had a great advantage to satisfy the requirements of industry. The local-bus clock of PCI is asynchronous with the PCI-bus clock. Whereas the PCI runs at 33 MHz with the burst transmission speed of 132 MB/s, the Local end can support the maximum clock frequency of 60MHz with the burst transfer speed of 200 MB/s. It has performance features such as two programmable FIFOs for zero wait state burst operation, supporting both multiplexed and non-multiplexed 8-, 16- or 32-bit address/data protocol.

In manufacturing machine, it is essential to provide a shared memory to access among processors in order to guarantee high-speed data transmission. As a result, Dual-Port RAMis showed high performance to complete this task, which handles the data transmission between host and DSP quickly and certainly. It has dual-ported memory cells which allow simultaneous access of the same memory location, expandable data bus to 72 bits and onchip arbitration logic.

To deal with the input of servo driver, Digital-Analog-Converter is used to convert digital control signal to analogy control signal, then, transfer it to servo driver. The design selects a quad, serial input, 12-bit DAC. The voltage output ranges from −10 V to +10 V.

SOFTWARE DESIGN OF MOTION CONTROLLER

Data Exchange

Elements of software design in this motion controller include library, firmware and windows-based application. Firmware is implemented on-board's DSP for motion profile generator, control algorithm to follow and reach target position. It also copes with control software to communicate with peripheral device through environment variable appropriately. Library acts as the interface of firmware and system software to perform the desired functions by providing optimal command functions. Besides, the purpose of windows-based application is to support user with tuning parameters interface and receiving the feedback signal to display in graph intuitively.

As shown in Figure 3, software data flow of the communication performs between controller and system software through DPRAM. In system software of host, the library data structure gets the possibility to communicate with firmware. Consequently, firmware and RAM memory use the same data structure in the operation. Most of data in firmware reads through DPRAM.

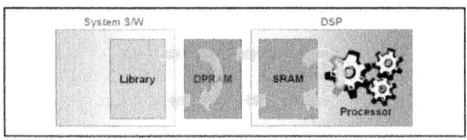

Figure 3: Data exchange in proposed motion controller.

If it needs to set the data or to perform the command via library, the corresponding data is written to DPRAM data and again copied to RAM memory for actual operation to use. When the operation completes, data in RAM memory is copied to DPRAM and feedback to system software.

Motion Algorithms

If motion controller is the key components of motion system, then control strategy is the key technology of motion controller. To drive the motor smoothly, S-curve profile generator algorithm is programmed in DSP to define the trajectory. In Figure 4, full S-curve profile is divided in seven segments including the values of jerk, acceleration, velocity and position.

For period $0 \leq \tau = \tau_1 \leq t_1$:

$$J(\tau) = J_{m1} \tag{1}$$

$$A(\tau) = J_{m1}\tau_1 \tag{2}$$

$$V(\tau) = \frac{1}{2}J_{m1}\tau_1^2 + V_0 \tag{3}$$

$$D(\tau) = \frac{1}{6}J_{m1}\tau_1^3 + V_0\tau_1 + D_0 \tag{4}$$

For period $t_1 < \tau = \tau_2 \leq t_2$:

$$J(\tau) = 0 \tag{5}$$

$$A(\tau) = A_m \tag{6}$$

$$V(\tau) = A_m\tau_2 + V_1 \tag{7}$$

$$D(\tau) = \frac{1}{2}A_m\tau_2^2 + V_1\tau_2 + D_1 \tag{8}$$

For period $t_2 < \tau = \tau_3 \leq t_3$:

$$J(\tau) = J_{m3} \tag{9}$$

$$A(\tau) = A_m - J_{m3}\tau_3 \tag{10}$$

$$V(\tau) = A_m\tau_3 - \frac{1}{2}J_{m3}\tau_3^2 + V_2 \tag{11}$$

$$D(\tau) = \frac{1}{2}A_m\tau_3^2 - \frac{1}{6}J_{m3}\tau_3^3 + V_2\tau_3 + D_2$$

(12)

For period $t_3 < \tau = \tau_4 \le t_4$:

$$J(\tau) = 0$$

(13)

$$A(\tau) = 0$$

(14)

$$V(\tau) = V_3$$

(15)

$$D(\tau) = V_3\tau_4 + D_3$$

(16)

For period $t_4 < \tau = \tau_5 \le t_5$:

$$J(\tau) = J_{ms}$$

(17)

$$A(\tau) = -J_{ms}\tau_5$$

(18)

$$V(\tau) = -\frac{1}{2}J_{ms}\tau_5^2 + V_4$$

(19)

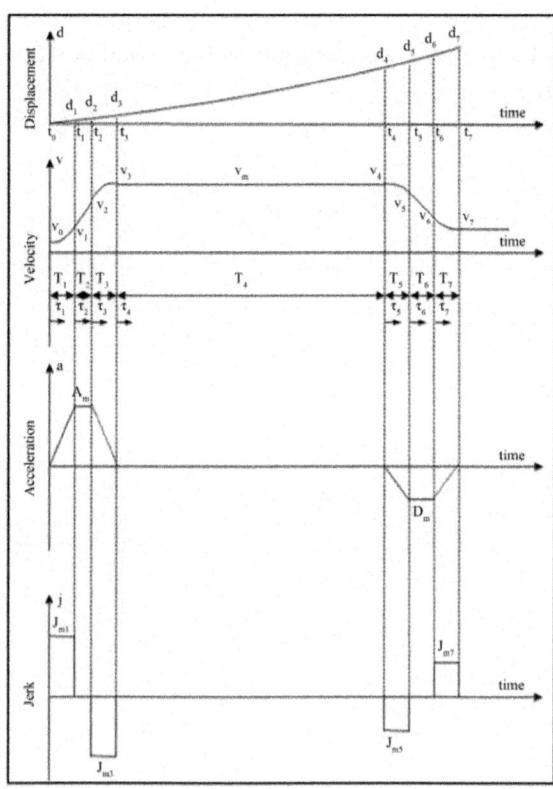

Figure 4: Full S-curve profile.

$$D(\tau) = -\frac{1}{6}J_{ms}\tau_5^3 + V_4\tau_5 + D_4 \tag{20}$$

For period $t_5 < \tau = \tau_6 \le t_6$:

$$J(\tau) = 0 \tag{21}$$

$$A(\tau) = -D_m \tag{22}$$

$$V(\tau) = -D_m\tau_6 + V_s \tag{23}$$

$$D(\tau) = -\frac{1}{2}D_m\tau_6^2 + V_s\tau_6 + D_s \tag{24}$$

For period $t_6 < \tau = \tau_7 \le t_7$:

$$J(\tau) = J_{m7} \tag{25}$$

$$A(\tau) = -D_m + J_{m7}\tau_7 \tag{26}$$

$$V(\tau) = -D_m\tau_7 + \frac{1}{2}J_{m7}\tau_7^2 + V_6 \tag{27}$$

$$D(\tau) = -\frac{1}{2}D_m\tau_7^2 + \frac{1}{6}J_{m7}\tau_7^3 + V_6\tau_7 + D_6 \tag{28}$$

where $A_x = A(t_x)$; $V_x = V(t_x)$ and $D_x = D(t_x)$; $x = 1,2,\cdots,7$; $A_m = \max(A)$; $D_m = \max(D)$

Once, the time duration in each segment is obtained from above equations. S-curve profile drives the servo motor to target position smoothly.

In industrial actuators, nonlinear and uncertain factors often exist throughout the operation. To reach the achievement of precision motion control, a system's nonlinearity must be eliminated. Many traditional approaches such as PID control scheme were researched. Nonetheless, conventional PID controller, adjusted using Ziegler-Nichols or trial-and-error technique, can usually meet poor performance requirements. But rising time and overshoot seem difficult to modify to be obtained the good result [21]. In order to have a better estimation of the system response, easier-tuning-topology, considerably separate overshoot and rising time, an alternative controller is needed. As a result, Proportional-Integral-Velocity control scheme is recommended. Nevertheless, with only a feedback control scheme, the tracking error due to servo lag phenomena still occurs. One of the existing solutions is to add the feed-forward controller into the main control scheme to enhance tracking performance. For this kind of design, the feed-forward control scheme is independent in case the values of parameters in the position loop controller change.

The overall control scheme of motion controller is described in Figure 5. The proposed controller includes a position loop with velocity loop. In this design, the error of position multiplied by K_p becomes a velocity correction command. The term K_V is used to estimate velocity of system. Later, the integral factor K_I is applied directly on the velocity error. Finally, both velocity and acceleration feed-forward unit are added to greatly minimize the tracking error.

K_P, K_I, K_V: Gains of proportional, integral and velocity loop respectively.

J_s, b, K_t, T_d, θ: Total inertia, damping, torque constant, disturbance torque, angular displacement of servo motor.

$\hat{K}_t$: Estimated torque constant of servo motor.

K_{vff}, K_{aff}: Velocity, acceleration feed-forward gain.

Position, velocity and acceleration reference are from motion profile generator.

The output torque control is presented:

$$u = u_{FF} + u_{PIV} \tag{29}$$

u_{FF} and u_{PIV} are torque control of feed-forward and PIV control scheme correspondingly.

Control Flow

The structure of control software is designed to obtain the high performance systematically. As shown in Figure 6, the main routine of motion controller is introduced. Firstly, system control registers are initialized. Then, several sections in DPRAM store the default values such as firmware version in index section, motion coefficients in motion section, sensor status in supervisor section. At the same time, RAM memory is activated for internal variables. The setting hardware and interrupt are completed before the end process of initialization. In the next task, the program stays in a loop named as Command Manager. This sub-function will manage the whole operations of all axes in motion system continuously.

The content of Command Manager is demonstrated in Figure 7. It can be seen clearly that controller spends a lot of time in Control Loop. During this time, the controller collects sensor signals and checks error conditions of amplifier and encoder. The Control Loop also manages the supervisor command executions such as motion commands and global commands. In Channel Control, several statuses of sending or receiving channel are defined to ensure the continuous communication between host and motion controller. In the case of ready status in channel, host sends the motion command to start computing path planning.

When motion is done or meets error, the controller will send interrupt to host to notify the condition of motion system. Then, data is updated through DPRAM to demonstrate the motion parameters in window-based application.

EXPERIMENTAL IMPLEMENTATION

In this section, experiment is set up on DSP-based motion control system in Figure 8. The host computer is Industrial PC with Windows OS.

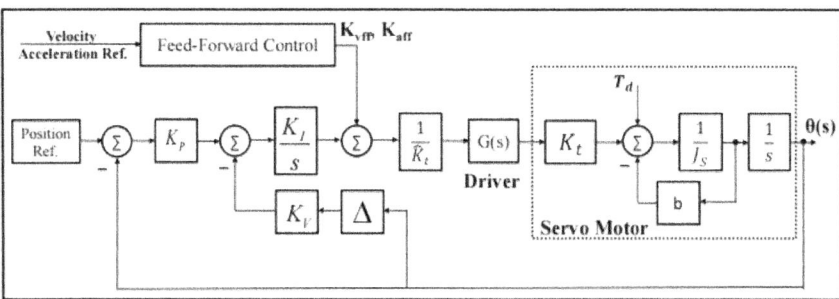

Figure 5. Block diagram of proposed motion controller.

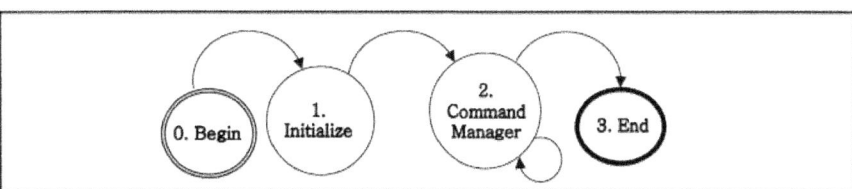

Figure 6. Main program flowchart.

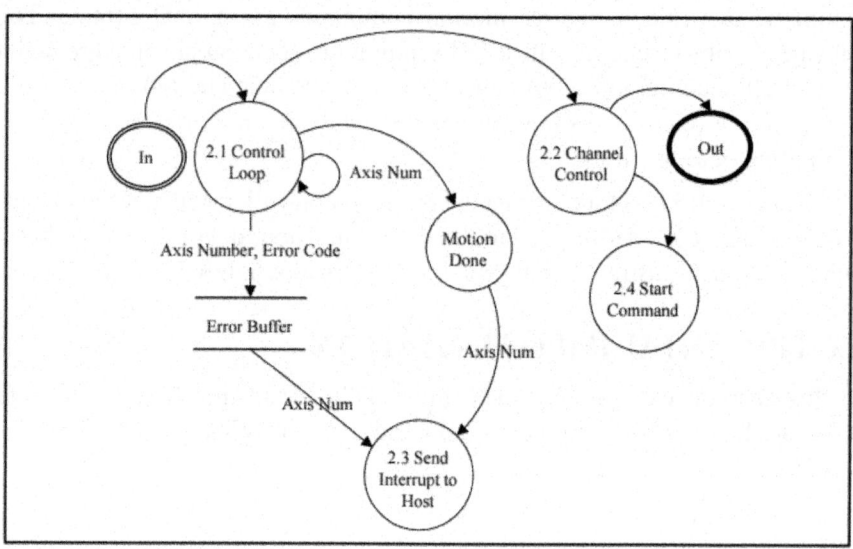

Figure 7. Flowchart of command manager.

The proposed motion controller can interface with user through Windows-based application. The AC servo motor SGMJV-02ADA21 which has an incremental encoder with a resolution of 2048 pulses/revolution is driven by servo driver SGDV-1R6A01B. All of them are produced by YASKAWA Company.

To verify the fast, smooth and accurate response of motion system, the proposed motion controller is programmed in industrial mode operations: acceleration motion, blending motion and step motion. The result data includes command and actual velocity, torque effort. All of these have the same control gains of motion controller, $K_p = 2.5$, $K_I = 0.02$, $K_V = 50$ and $K_{VFF} = 30$. In Figure 9, experimental performance is achieved with full S-curve motion planning strategy. The result indicates that the controller brings the actual profile to the specified profile smoothly. In blending motion test of Figure 10, a simple strategy to blend two segments together by shifting the speed segment along the time domain. Although actual velocity varies in several desired values during the process of motion, the torque is observed to be stable; hence, smoothness and accuracy are still guaranteed to obtain high performance. As shown in Figure 11, the motor closely tracks the pre-defined trajectory even at short time.

Figure 8. Experimental motion system.

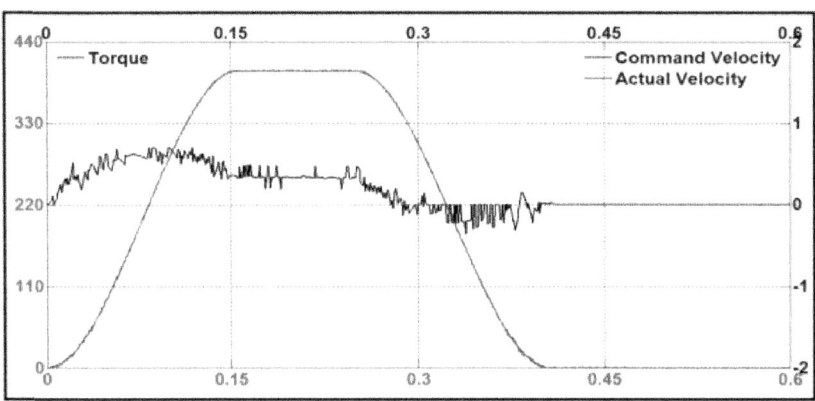

Figure 9. Performance of proposed motion controller in acceleration motion mode.

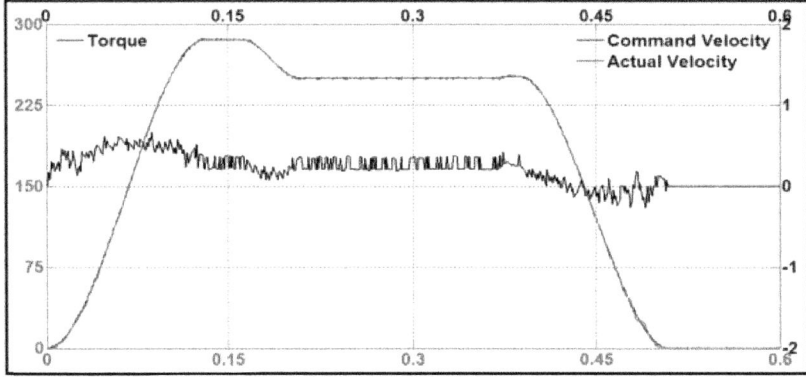

Figure 10. Performance of proposed motion controller in blending motion mode.

Clearly, the response is fast and satisfies the requirements of real-time control. Moreover, the operation of motion system is promised to be stabilized in most of executing duration.

It can be seen that the proposed motion controller has a superior performance and highly machining accuracy. Servo motor is sufficiently driven to track desired trajectory regardless of the types of motion modes. The programming implementation is based on DSP control board and C/C++ language. This environment permits most of motion control engineers to develop the extensive applications not only in the laboratory, but also in practical factory.

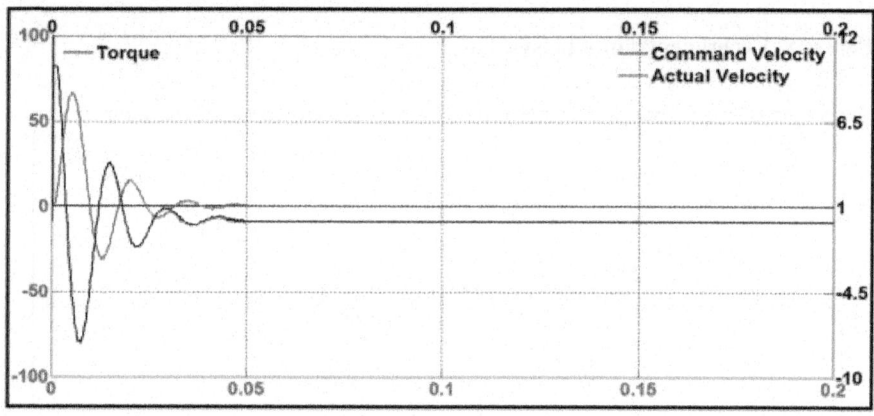

Figure 11. Performance of proposed motion controller in step motion mode.

In this way, it is suggested that the developed motion controller is significantly suitable for manufacturing machine.

CONCLUSION

In this paper, a feasible and effective control scheme has been proposed for closed-loop motion system. The controller implemented feed-forward PIV strategy which combines velocity/acceleration feed-forward control unit with a velocity loop. The results in experiment prove the smooth profile, highly tracking accuracy and realtime performance. Thus, this controller is suggested to be applicable in industrial precision closed-loop motion control system like CNC machine, robot control.

FUNDING

This Work was supported by Dong-eui University Foundation Grant (2011).

REFERENCES

1. Jun, W.X., Dong, S.S., Xue, T.Z. and Yang, H.B. (2010) The Remote Monitor and Control System for Motion Controller Based on the Ethernet. International Conference on Intelligent System Design and Engineering Application, 2, 717-719.http://dx.doi.org/10.1109/ISDEA.2010.215

2. Sung, M.Y., Kim, K.H., Jin, H.W. and Kim, T.H. (2011) An EtherCAT-Based Motor Drive for High Precision Motion Systems. International Conference on Industrial Informatics, Caparica, 26-29 July 2011, 163-168. http://dx.doi.org/10.1109/INDIN.2011.6034856

3. Ruili, C. and Jun, H. (2010) Developing and Research on Motion Controller Based on LM628. International Conference on E-Product E-Service and E-Entertainment, Henan, 7-9 November 2010, 1-3. http://dx.doi.org/10.1109/ICEEE.2010.5661605

4. Qiang, Y., Jing, F., Hou, Z., Yang, S. and Liu, Y. (2012) Experimental Validation of a Trajectory Planning Method withContinuous Acceleration Implemented on DSP-Based Motion Controller. Proceedings of the 10th World Congress on Intelligent Control and Automation, Beijing, 7-9 November 2010, 3326-3330.http://dx.doi.org/10.1109/WCICA.2012.6358447

5. Qiang, Y., Jing, F., Hou, Z. and Li, E. (2010) A Design of Multi-Axis Motion Controller for Welding Robot Based on DSP. International Conference on Advanced Technology of Design and Manufacture, Beijing, 23-25 November 2010, 383-387.http://dx.doi.org/10.1049/cp.2010.1328

6. Komin, U. and Tanta-ngai, K. (2011) DSP-Based Motion Controller Development for Milling Machine. Proceedings of SICE Annual Conference, Tokyo, 13-18 September 2011, 849-853.

7. Lv, C.H., Guo L.J. and Li J.D. (2012) Research of Motion Control System Based on PCI-1243.Third International Conference on Digital Manufacturing and Automation, 31 July - 2 August 2012, Guilin, 662-665. http://dx.doi.org/10.1109/ICDMA.2012.157

8. Arif, S., Iqbal, J. and Munawar, S. (2012) Design of Embedded Motion Control System based on Modified Fuzzy Logic Controller for Intelligent Cruise Controlled Vehicles. International Conference on Robotics and Artificial Intelligence, Rawalpindi, 22-23 October 2012, 19-25. http://dx.doi.org/10.1109/ICRAI.2012.6413421

9. Ling, X., Li, Q., Wang, T., Dong, J., Tang, Z. and Ding, Y. (2011) Research and Design of Motion Controller for CNC Based on Fuzzy PID Algorithm with Feed-Forward Control. International Conference on

Control, Automation and Systems Engineering, Singapore, 30-31 July 2011, 1-4. http://dx.doi.org/10.1109/ICCASE.2011.5997519

10. Zhou, L., Li, J., Sheng, J., Cao, J. and Li, Z. (2010) Closed-Loop Identification for Motion Control System. Proceedings of the 8th World Congress on Intelligent Control and Automation, Jinan, 7-9 July 2010, 477-480.http://dx.doi.org/10.1109/WCICA.2010.5553795

11. Rubaai, A. and Young, P. (2010) DSP-Based Fuzzy Neural Network PI/PD-like Fuzzy Controller for Motion Controls and Drives. IEEE Industry Applications Society Annual Meeting, Houston, 3-7 October 2010, 1-8. http://dx.doi.org/10.1109/IAS.2010.5614081

12. Rubaai, A. and Young, P. (2011)Hardware/Software Implementation of PI/PD-like Fuzzy Controller for High Performance Motor Drives. IEEE Industry Applications Society Annual Meeting, Orlando, 9-13 October 2011, 1-7. http://dx.doi.org/10.1109/IAS.2011.6074341

13. Dey, C., Mudi, R.K. and Mitra, P. (2012) A Self-Tuning Fuzzy PID Controller with Real-Time Implementationon A Position Control System. Third International Conference on Emerging Applications of Information Technology, Kolkata, 30 November - 1 December 2012, 32-35. http://dx.doi.org/10.1109/EAIT.2012.6407855

14. Rubaai, A. and Jerry, J. (2010) dSPACE DSP-Based Real-Time Implementation of Fuzzy Switching Bang-Bang Controller for Automation and Appliance Industry. IEEE Industry Applications Society Annual Meeting, Houston, 3-7 October 2010, 1-8.http://dx.doi.org/10.1109/IAS.2010.5614479

15. Rubaai, A. and Jerry, J. (2011) Hybrid Fuzzy Bang-Bang Mode Controller for Electric Motor Drives Applications. IEEE Industry Applications Society Annual Meeting, Orlando, 9-13 October 2011, 1-8. http://dx.doi.org/10.1109/IAS.2011.6074336

16. Meza, J. L., Santibanez, V., Soto, R. and Llama, M.A. (2012) Fuzzy Self-Tuning PID Semiglobal Regulator for Robot Manipulators.IEEE Transactions on Industrial Electronics, 59, 2709-2717. http://dx.doi.org/10.1109/TIE.2011.2168789

17. Chiew, T.H., Jamaludin, Z., Bani Hashim, A.Y., Rafan, N.A. and Abdullah, L. (2013) Identification of Friction Models for Precise Positioning System in Machine Tools. Procedia Engineering, 53, 569-578. http://dx.doi.org/10.1016/j.proeng.2013.02.073

18. Lin, W.F., Yeh, S.S. and Sun, J.T. (2011) Friction Compensation Design for Velocity-Controlled Feed Drive Motions of CNC Machines. 9th World Congress on Intelligent Control and Automation, Taipei, 21-25

June 2011, 182-187.http://dx.doi.org/10.1109/WCICA.2011.5970725

19.　Huang, W.S., Liu, C.W., Hsu, P.L. and Yeh, S.S. (2010) Precision Control and Compensation of Servomotors and Machine Tools via the Disturbance Observer. IEEE Transactions on Industrial Electronics, 57, 420-429.http://dx.doi.org/10.1109/TIE.2009.2034178

20.　Rijlaarsdam, D., Nuij, P., Schoukens, J. and Steinbuch, M. (2012) Frequency Domain Based Nonlinear Feed-Forward Control Design for Friction Compensation. Mechanical Systems and Signal Processing, 27, 551-562.http://dx.doi.org/10.1016/j.ymssp.2011.08.008

21.　OH (2001) Motion System Handbook. Power Transmission Design, Penton Media Inc., Cleveland.

Chapter 6

MODEL-DRIVEN DEVELOPMENT OF AUTOMATION AND CONTROL APPLICATIONS: MODELING AND SIMULATION OF CONTROL SEQUENCES

Timo Vepsäläinen and Seppo Kuikka

Department of Automation Science and Engineering, Tampere University of Technology, P.O. Box 692, Korkeakoulunkatu 3, 33101 Tampere, Finland

ABSTRACT

The scope and responsibilities of control applications are increasing due to, for example, the emergence of industrial internet. To meet the challenge, model-driven development techniques have been in active research in the application domain. Simulations that have been traditionally used in the domain, however, have not yet been sufficiently integrated to model-driven control application development. In this paper, a model-driven development process that includes support for design-time simulations is complemented with support for simulating sequential control functions. The approach is implemented with open source tools and demonstrated by creating and simulating a control system model in closed-loop with a large and complex model of a paper industry process.

INTRODUCTION

Model-driven development (MDD) is a system and software development methodology that emphasizes the use of models during the development work. In MDD, models conform to modeling languages that have formal metamodels, for example, unified modeling language (UML). In addition to manual development work, models can be processed with model transformations that revise existing and create new, refined models. The use of transformations may automate error-prone tasks such as importing information to models from preceding development phases and tools. Design models can be used for generating code or to analyze the developed systems. Automated model checks may reveal problems and inconsistencies in models and between phase products.

The mentioned benefits of MDD are related to development tasks that are repetitive and simple enough to be treated with preprogrammed rules. However, MDD has not been able to, and probably cannot, automate all the complex tasks in system and software development. Demanding design decisions over alternative solutions to achieve (sometimes informal) objectives and product characteristics need to be made by professional developers. However, although genuine design decisions cannot be automated, developers do not always have to rely solely on their experience. For example, simulation is a technique that has been traditionally used in the domain within control algorithm development and control system testing.

Automation and control system development is also an application domain in which the use of MDD techniques has been researched extensively during recent years. However, despite the research activities and the tradition of using simulations, ability to simulate early software design models has not yet been sufficiently addressed in the domain.

In their previous work, the authors have developed a simulator integration [1] to the tool-supported Aukoton MDD process [2] for automation and control applications. The approach is based on UML Automation Profile (UML AP) [3]. It enables modeling and simulation of cyclically executed control functions including feedback and binary control as well as interlocks (interlocks are used in control systems to protect the controlled processes from causing harm to themselves or personnel, e.g., by forcing actuators to safe states based on measured states of the processes).

The simulation support is intended to be usable during both platform independent and platform specific modeling phases of the development process; see Figure 1. During the platform independent phase, it is possible to, for example, evaluate alternative control approaches, structures, and interlocks. During the platform specific phase, the approach enables the evaluation of platform specific functions, tunings, and predicted overall performance of the system. Technically the approach utilizes model-in-the-loop simulations so that UML AP control system models are transformed to ModelicaML models. In this paper the approach is complemented by enabling simulation of sequential control activities. The activities are modelled with Automation Sequences of UML AP and visualized with Automation Sequence Diagrams.

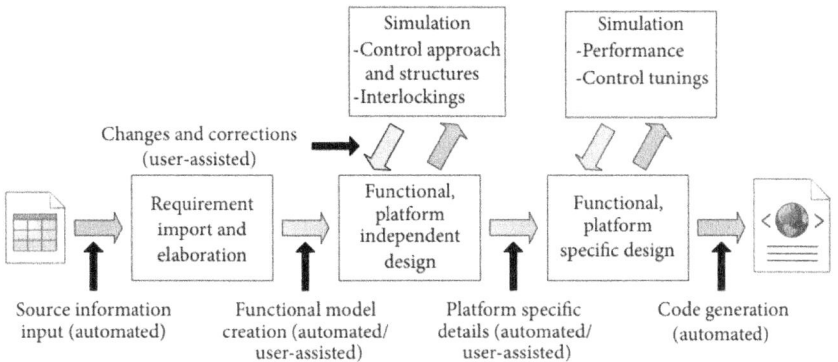

Figure 1: The MDD process with simulation extensions.

The contributions of this paper are as follows. The modeling notation is discussed in comparison to the well-known state machine notation of UML. An approach that enables the simulation of control sequences in a state-machine-like form is presented and implemented as a model transformation. The approach is integrated to the previous simulation integration. The approach is applied to batch control of a paper industry process.

The rest of this paper is organized as follows. Section 2 reviews work related to the use of MDD and simulations in the industrial control domain. In Section 3, the previous work is introduced briefly, which is necessary for understanding how the new work integrates to it. Section 4 discusses the use of control sequences in process industry, presents the UML AP approach to modeling control sequences, and presents the model transformation to create simulation models. In Section 5, before discussion and conclusions, the approach is applied to an illustrative pulp batch processing system.

RELATED WORK

The use of model-driven techniques has been researched extensively in the domain of industrial control during recent years. Modeling of requirements, architecture, and details of control applications has been seen as an important part of design processes and as a means to cope with the ever-increasing size and complexity of the applications. Many of the recent approaches have also integrated simulations to the development processes in order to be able to test early and concurrently to the development work.

In addition to industrial control, MDD with simulation features has been applied to control system development for automotive and other embedded applications. The general simulation approaches that can be applied when models are used for generating code include model-in-the-loop (MiL), software-in-the-loop (SiL), processor-in-the-loop (PiL), and hardware-in-the-loop (HiL) simulations [4]. The approaches differ in the control system configurations that are used to control plant models in closed-loop simulations. Examples on use of MiL, SiL, and HiL simulations in the embedded system domain include [5] that describes a general framework for and two examples of use of MiL simulation. A testing environment that uses SiL simulations is presented in [6]. HiL simulation and testing have been utilized, for example, in [7–9].

Another classification of simulation approaches is related to the amount of simulation engines. Simulation of a controlled system, with a model of a process to be controlled and a model of a control system, can be performed within a single simulation engine or as cosimulation. In cosimulation, the models are simulated within different but connected environments. This requires a mechanism to synchronize the simulations including their values and states. Commands and functions, for example, running, replaying, freezing, and loading states (see [10] for a list of basic simulation functions) must be replicated to all used engines.

The management and coupling of cosimulations have been recently addressed with FMI standard [11] and also with model based techniques [12]. However, the area of expertise of control application developers may still not be in simulation techniques. As a consequence, the use of a single simulation engine can be considered more recommendable.

In the industrial control domain, MAGICS approach for MDD of industrial process control software is presented in [13]. As a modeling notation the approach utilizes ProcGraph that has been implemented on the Eclipse platform on top of Eclipse Modeling Framework (EMF). The approach utilizes several diagram types including Entity Diagrams (ED), State Transition Diagrams (STD), and State Dependency Diagrams (SDD), of which STD is suitable for modeling sequential behavior. The approach enables the generation of executables but does not address simulations.

The FLEXICON project studied the integration of Commercial-Off-The-Shelf tools, including MATLAB/Simulink and ISaGraf, to support the development of control applications for marine, automotive, and aerospace systems [14]. The approach uses cosimulation, which is enabled by DSS (data delivery service) middleware between the tools.

Vyatkin et al. [15] developed a model-integrated design framework for designing and validating industrial automation systems. It is based on the Intelligent Mechatronic Component (IMC) concept and the use of IEC 61499 architecture. New systems are developed from IMCs that are integrated together and with their models enable formal verification, closed-loop MiL simulation of IEC 61499 models and code deployment.

The approach of the MEDEIA project [16] builds on use of several model types as well as bidirectional model transformations. The process supports the use of closed-loop MiL simulations which are based on use of an IEC 61499 environment. Simulation models of the process parts are in the approach defined with either timed state charts or external behavior descriptions (external simulation tools).

The abovementioned standard, IEC 61499 [17], is a specification and modeling language for industrial control applications. It extends the function block concept of another IEC programming language, IEC 61131-3 [18], with event driven execution and support for distribution of applications. With an appropriate tool support, IEC 61499 models can also be used for simulation purposes.

The simulation approach in [19] is based on mixing real control hardware with simulated one while simulating the plant in another (SIMBA 3D) environment. The benefit of the cosimulation approach is the ability to test early, by executing already implemented parts and simulating the rest.

The simulation approach closest to our work has been recently presented in [20]. In a manner similar to [15, 16] IEC 61499 is used for simulation purposes also in [20]. Model transformations are used for creating IEC 61499 plant models from MATLAB/Simulink plant models to obtain closed-loop behavior within a single (MiL) simulation environment.

Difference from the work to be presented, the referred simulation approaches in the domain ([14–16, 19, 20]) do not address modeling and simulation of sequential control separately from, for example, stabilizing feedback control. In [13] the sequential control aspect is addressed with respect to modeling. However, simulation of the models is not suggested. The use of simulations can thus be assumed to be possible no earlier than after code generation.

On the other hand, the referred development approaches that support simulations rely either on cosimulation ([14, 19]) or use of IEC 61499 as a simulation language ([15, 16, 20]). The use of IEC 61499 to simulations was not a viable alternative in this work to be presented because it is not used in the MDD process [2] that is extended with simulations. On the other hand, UML AP models that are used in the development process are not simulatable as such. The use of cosimulation would thus have either required a transformation to a simulatable form or delayed simulations to simulating plant models with produced executables. These reasons, however, apply to a number of MDD processes with nonsimulatable modeling languages such as UML and UML profiles.

Other works related to sequential control with model-based characteristics include [21] that presents an approach to transform Grafcet [22] models to Mealy machines for testing purposes (Grafcet is a conventional means to specify control sequences). Execution semantics of Sequential Function Charts (SFC) [18] have been addressed in [23]. The SFC notation is part of IEC 61131-3 [18] and based on the earlier version Grafcet.

In the simulation approach to be presented, the target simulation language is Modelica [24] with Modelica Modeling Language (ModelicaML) [25] as an intermediate language. Modelica is an object-oriented, equation based simulation language. The basic concepts of it are simulation classes that contain properties, equations, and connectors. Similarly to classes of object-oriented programming languages, Modelica classes can inherit properties (and equations) of parent classes. Simulatable Modelica models consist of instances of the classes that are connected together with their connectors. ModelicaML [25], on the other hand, is a UML profile for Modelica. It consists of stereotypes and tagged values that correspond to the key words and features of Modelica and enable modeling of Modelica models with UML tools. ModelicaML models are not simulatable as such but can be transformed to simulatable Modelica form with OpenModelica tools [26]. For simulating Modelica models there are both open source (e.g., OpenModelica [26]) and commercial tools (e.g., Dymola [27]) available.

The ModelicaML (profile) implementation uses UML2 plugins on the Eclipse platform which are built on Eclipse Modeling Framework (EMF) implementation of OMG Meta Object Facility (MOF). The ModelicaML implementation is thus technically similar to the UML AP implementation [28] that is used in this work for control system modeling. It has been implemented by extending UML2 and Topcased SysML metamodel with EMF. This similar background of the tools enables implementing the transformation from UML AP to ModelicaML using standardized QVT (Query/View/Transformation) languages [29] and their open source implementations on the Eclipse platform.

On Simulating Control Application Models

The objective of integrating simulations to MDD for automation and control applications is to support design-time quality assurance activities. It should be possible to compare alternative control approaches and structures and tunings as well as interlocks. Design flaws should be found and corrected as early as possible and to the extent possible so that they would not affect adversely subsequent design phases. By enabling simulation of design-time models, it could be also possible to obtain at least part of the general benefits of simulations before implementation of the applications. Such general benefits include, for example, improvements to the design, development, and validation of the control programs, as reported in [10].

Without specific support for sequential control, the approach to create simulation models from UML AP models has been presented in [1]. In UML AP, the modeling concepts for functional modeling are automation functions (AFs) that have been divided to a hierarchy of measurements, actuations, controls, and interlocks. Measurement and actuation AFs are interfaced with sensors and actuators of the controlled processes while performing conversions of signals to and from engineering units. Control AFs perform computation of control signals according to control algorithms. The purpose of interlock AFs is to compute releasing and locking signals for actuators and devices. AFs interchange signals and information with ports.

The transformation for simulating functional UML AP models (that consist of AFs) creates and appends simulation counterparts of the AFs to Modelica plant simulation models. For platform independent AFs, the transformation utilizes a library of predefined simulation counterparts (classes) of them. To support platform and vendor specific AFs, the transformation is capable to utilize external libraries of simulation classes. To support application specific AFs, for example, interlocks that require tailoring for each application, the transformation is capable to create simulation classes based on logic diagram descriptions of AFs [1]. The process described in this paper is an equivalent approach to create new simulation classes but based on Automation Sequence Diagrams instead of logic diagrams.

The decision to use model transformations in this manner was made because UML AP models, as they are used in the tool, are not simulatable as such. Transforming plant models to the control application models would not have enabled closed-loop simulation, for example, in [16, 20] in which (IEC 61499) models were simulatable. In a similar manner, use of cosimulation as in [14, 19] would have required transformation to a simulatable form before applying cosimulation. Additionally, the cosimulation approach would require additional work; see Section 2, related to, for example, coupling simulations skills that not all control application developers can be assumed to have. The approaches to obtain closed-loop simulations by transforming plant models, by transforming control application models, and by using cosimulation have also been recently compared in [30].

An example structure of a plant model before and after executing the transformation that appends the control application specific parts to it is illustrated in Figure 2. Before executing the transformation, the model contains simulation class definitions of the parts of the plant and a description of how the interconnected instances of the classes form the system model. This part of the model, referred to as the original process model, is circled with blue, dashed line. The transformation (1) copies and creates new simulation class definitions based on the control system model, (2) creates instances of the classes according to the control system model, and (3) couples the required instances of the classes to the original model. In the figure, the newly created parts of the model are circled with red, dashed line.

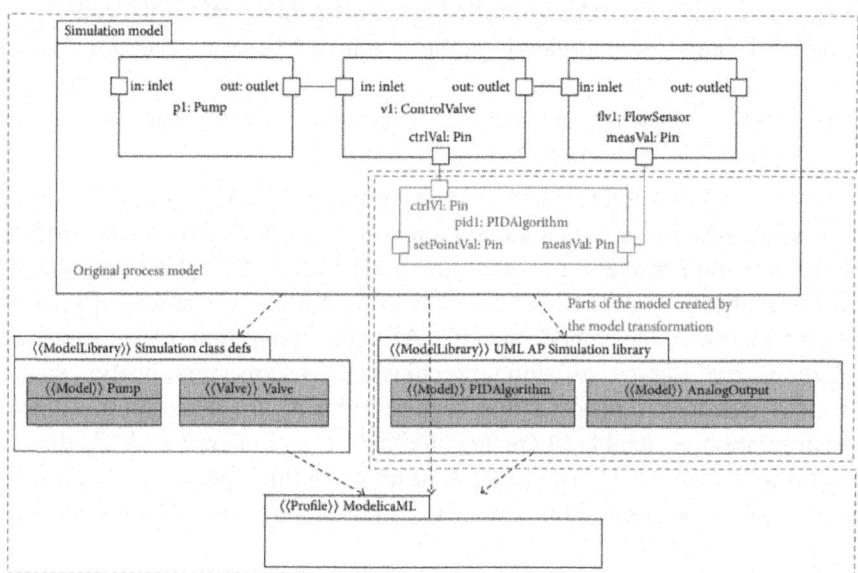

Figure 2: The transformation adds the control system specific parts to an existing model of the physical process.

MODELING AND SIMULATION OF CONTROL SEQUENCES

Control sequences are needed by process industries to perform start-ups of complex processes, for example, power plants or paper machines and to drive them to their designed operating states. In a similar manner, shutting down a process in a controlled and energy efficient manner may require changing set-points of process variables and shutting down devices and sub-systems in a specific order. On the other hand, batch processes constitute a challenging part of industrial processes. In batch processes production of the end products may require, for example, addition of source materials and substances according to time constraints and achievement of defined process states, for example, temperatures and concentrations.

The UML AP approach to modeling sequential control is based on (automation) sequences that have been developed to enable a SFC-conformant modeling notation within UML AP models. Sequences are modelled with a domain specific, new diagram type, Automation Sequence Diagram (ASD). Graphically the ASD notation resembles both the state machine and activity modeling notations of UML.

Description of the Modeling Notation

Sequences that are described in ASDs consist of Steps that are basic procedural elements in the approach (e.g., upper level batch recipe steps or device level controls). Similar to states of UML state machines, Steps contain Entry, Step, and Exit Activities that are executed when arriving to the Step, during the Step and when exiting the Step, respectively. In addition, Steps may reference other Sequences that can be defined with other ADSs. This is an equivalent characteristic to composite states of UML. Containing activities and referencing a sub-Sequence are exclusive alternatives for a Step. In addition to basic Steps, Sequences may contain Allocations. Allocations are intended for reserving process items and devices for the Sequences that they appear in. When used, Allocations are next to initial Steps in the Sequences.

The execution order of Steps within a Sequence is determined by Transitions that may contain different kinds of conditions that control when a Transition is fired. First, the condition can be a Boolean condition that explicitly specifies a Boolean valued condition based on, for example, values of the variables of the AF that contains the Sequence. Secondly, a condition can be a timeout condition specifying how long the Transition must wait after the execution of the previous Step is finished. Additionally, the Transition can be a one shot Transition which is fired immediately after the previous Step has been executed.

In addition to Steps, Allocations, and Transitions, Sequences contain initial and final as well as fork and join Steps. They can be used in a similar manner that the corresponding pseudostates of UML state machines, that is, to control the execution of Sequences. Use of initial and final Steps is also a necessity in each Sequence because whether a transition may occur from a Step to another is not always dependent (only) on the conditions of the Transitions.

Consider, for example, the two example diagrams in Figure 3. The figure also illustrates the graphical representation of initial and final Steps, (basic) Steps, Steps that reference sub-Sequences, forks, joins, and Allocations. In the Sequence at the left-hand side, all the Steps reference sub-Sequences that consist of Steps and possibly other sub-Sequences. For example, the WLF (White Liquor Fill) Sequence in the right-hand side diagram is referenced from the third Step in the left-hand side diagram.

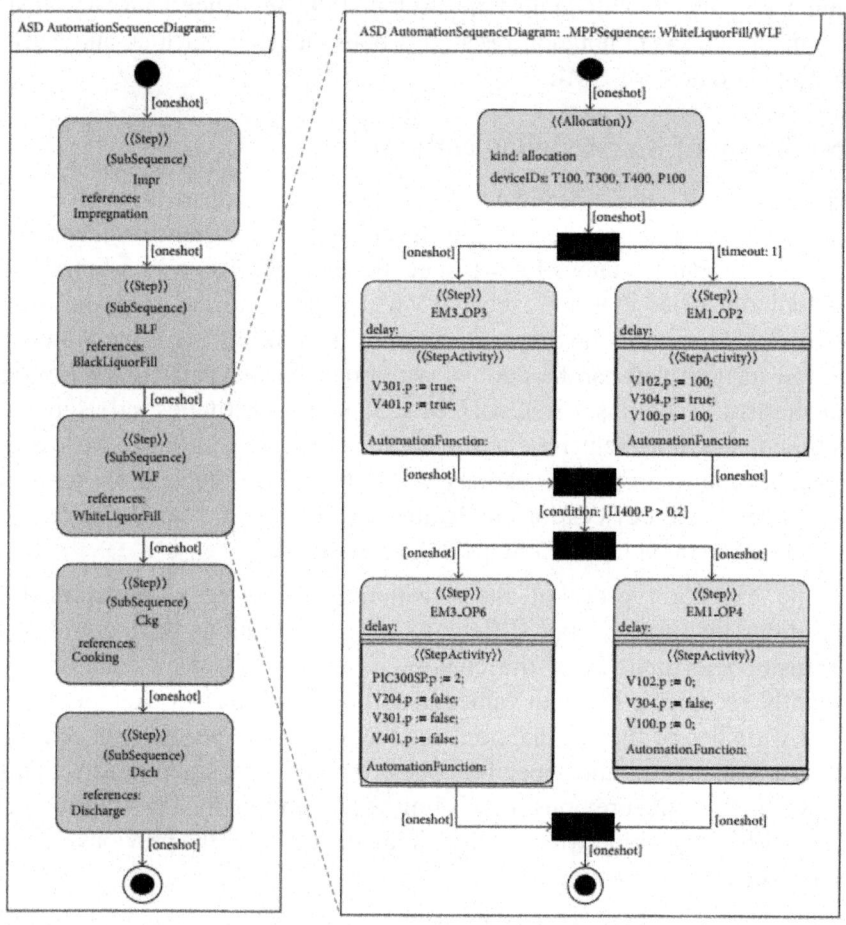

Figure 3: Automation sequence diagrams illustrating a sequence and a referenced subsequence of it.

Because the Transitions in the Sequence at the left-hand side are one shot Transitions, it is obvious that whether a Transition can fire is also dependent on the completion of the referenced Sequences. Referenced Sequences need to have performed their control activities in a similar manner as in SFCs [18], which is a domain specific notation based on which the ASD notation has been developed. For a Transition to be fired from a Step referencing a sub-Sequence, the referenced Sequence must have reached its final Step. This is a clear semantic difference of the notation in comparison to UML state machines.

Some other obvious differences to UML state machines are also visible in Figure 3. The first Step in the Sequence on the right-hand side is an Allocation. In the example, the allocated process parts are (tanks) T100, T300, and T400 as well as (pump) P100. (In UML state machine diagrams, there are no similar concepts.) After the first of the fork Steps, the transition condition on the right-hand side is of type timeout with value "1" indicating that the Transitions must wait 1 (sec) after the execution reaches the fork. In UML state machines the semantics of the timeouts is slightly different since the waiting time of UML AP Transitions starts from the completion time of the Step preceding the transition.

Lastly, as can be seen in the example Sequence at the right-hand side in Figure 3, Sequences may have several branches executing at the same time. In UML state machines, an analogous feature would be the possibility for a system to be in two (or more) states at the same time. This requires using composite states, each within a region of its own.

Another modeling notation of UML that the ASD notation resembles (both graphically and semantically) is the activity diagram notation that would enable concurrent Sequences of activities and explicit constraints on flows but no timing constraints. However, UML activities cannot be broken up to concepts corresponding to Entry, Step, and Exit Activities of Steps. Activity diagrams may additionally contain decision nodes for which there are no corresponding concepts in ASDs. Lastly, activity diagrams usually describe workflows of entire systems, whereas in UML AP Sequences are used to describe sequential behavior of individual AFs.

Because of the mentioned conceptual differences to the modeling notations of UML that have similar appearance, it was not possible to use directly research work that has been previously done to enable their simulation.

Model Transformation for Simulating Sequences

In general, Modelica is an equation based language so that the values of variables of Modelica models are determined by equations. However, in addition to the equations that apply all the time, the language includes an algorithmic concept for calculations in which statements are applied in an order. Algorithms are also the constructs of the language that the transformation uses for simulating the Sequences.

The simplified (hiding unnecessary details) metamodel of the ASD diagram type is presented in Figure 4. In the metamodel, Sequence is extended from the UML state machine. The Step and Allocation concepts are extensions of UML state. Entry, Step, and Exit Activities are extended from the UML activity concept and contained by Steps with metamodel properties of UML State (that are hidden from the figure). In addition to the concepts that are shown in the figure, ASDs may contain instances of the mentioned pseudostates of UML, namely, initial, join, and fork states (Steps) as well as final states. Transitions between Steps, Allocations, and pseudo Steps are modelled with a Transition concept that has been extended from UML Transition.

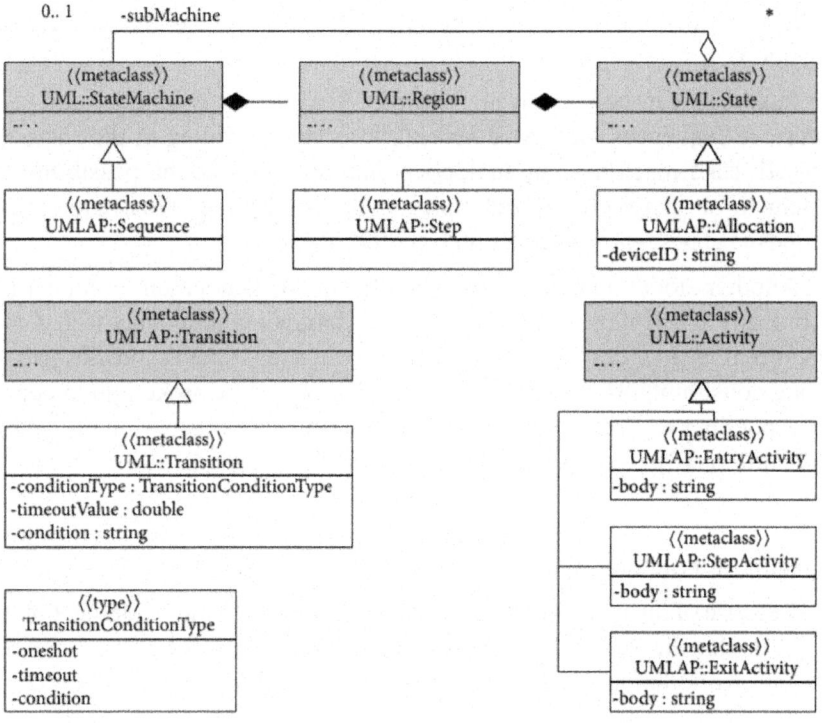

Figure 4: Simplified metamodel of the ASD diagram type with relations to the UML metamodel.

To simulate the behavior of Sequences, they are used as a basis for creating variables and algorithmic code. The systematically named variables are used to keep track of the execution, whereas the algorithmic code changes the values of the variables. The (Entry, Step, and Exit) Activity code of Steps is also included in the algorithms. The variables, which are created to be owned by a Modelica class that corresponds to the AF that owns the Sequence, are created as follows. For the sequence that an ASD represents, and, for each sub-Sequence that is referenced from the Steps of the Sequence, a Boolean variable with the same name than the name of the Sequence is created. These variables are used to indicate the execution of the Sequence being in the Sequence or sub-Sequence in question. In addition, exactly one UML OpaqueBehavior for each highest level Sequence is created to contain the algorithmic code to be generated.

For each Step in a Sequence, the transformation creates two variables. First is a Boolean variable with a name consisting of the name of the Sequence and the name of the Step. The second variable is an Integer variable with a name consisting of the name of the sequence, the name of the Step, and "Phase" literal. The Boolean variables are used to indicate the execution of the Sequence being in the Step in question, whereas the Integer variables keep track of which Activities (Entry, Step, or Exit) have been executed in a Step.

For (exactly one) initial Step in a Sequence, the transformation creates a Boolean variable with a name consisting of the name of the Sequence and "Initialized" literal. For a final Step in a Sequence, the transformation creates a Boolean variable with a name consisting of the name of the Sequence and the name of the final Step. These variables indicate whether or not the execution has reached the initial and final steps in question.

For each fork-to-join region the transformation creates a Boolean variable with a name that consists of the name of the fork, the name of the join, and "Region" literal. In addition, a Boolean variable is created for each branch going out from the fork and coming into the (exactly one) join. The names of these variables consist of the name of the fork (Step) and the number of the branch. The variables corresponding to the branches are used in guard conditions for exiting the join (Step), whereas the other variables are used to indicate the execution of the Sequence being in the fork-to-join region.

For Transitions, the transformation creates variables only if their transition condition is of type timeout. In this case, the name of the real valued variable consists of the name of the Sequence, the name of the Step from which the transition starts, and "Time" literal. These time variables keep track of completion times of the Steps that the Transitions exit from. Lastly, for the Allocations, the transformation generates a record (class) and a property for each individual device ID that becomes reserved in the Sequences owned by the AF. The mappings between UML AP and UML metamodel elements are also presented in Table 1.

Table 1: Mappings between UML AP and UML (ModelicaML) metamodel elements

Source model (UML AP)	Target model (UML with ModelicaML)		
Element	Model element	Element name	Element type
Sequence	Property	Seq. name	Boolean
	Opaque Behavior	Seq. name + "Algorithm"	—
(UML) Initial (pseudostate)	Property	Seq. name + "Initialized"	Boolean
Step	Property	Seq. name + Step name	Boolean
	Property	Seq. name + Step name + "Phase"	Integer
(UML) FinalState	Property	Seq. name + FinalState name	Boolean
(UML) Fork (pseudostate)	Property	Seq. name + Fork name + "Branch" + #	Boolean
(UML) Join (pseudostate)	Property	Fork name + Join name + "Region"	Boolean
Transition	Property	Seq. name + Step name + "Time"	Double
Allocation	Property	Seq. name + allocation name	Boolean
	Class	"Allocations"	—
	Property	Device ID	Integer

Some of the algorithmic constructs that are created based on the ASDs are illustrated in an example in Figure 5. First, a when-construct is created that is executed only once at the 7 start of the simulation ("when initial() then"). It sets all the Boolean phase variables (Steps, pseudo Step, Allocations, and sub-Sequences) to false. The Integer variables related to Steps are set to 0 to indicate that no activities have been performed. The Integer variables related to Allocations are also set to 0, to indicate that no allocations are active. The initialization code is created only for each highest level Sequence, not for referenced sub-Sequences.

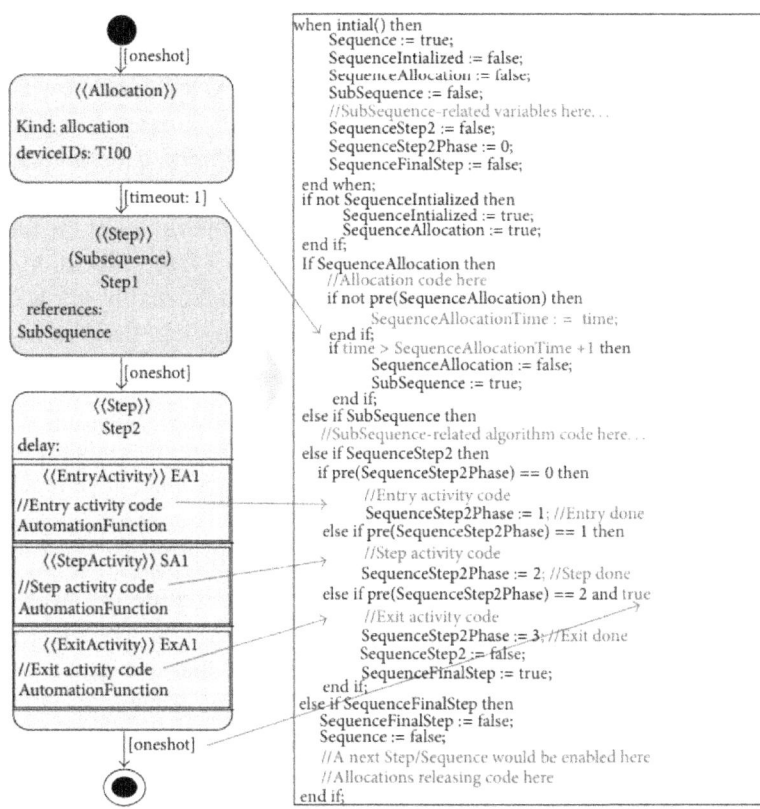

```
when intial() then
    Sequence := true;
    SequenceIntialized := false;
    SequenceAllocation := false;
    SubSequence := false;
    //SubSequence-related variables here...
    SequenceStep2 := false;
    SequenceStep2Phase := 0;
    SequenceFinalStep := false;
end when;
if not SequenceIntialized then
    SequenceIntialized := true;
    SequenceAllocation := true;
end if;
If SequenceAllocation then
    //Allocation code here
    if not pre(SequenceAllocation) then
        SequenceAllocationTime : = time;
    end if;
    if time > SequenceAllocationTime +1 then
        SequenceAllocation := false;
        SubSequence := true;
    end if;
else if SubSequence then
    //SubSequence-related algorithm code here...
else if SequenceStep2 then
    if pre(SequenceStep2Phase) == 0 then
        //Entry activity code
        SequenceStep2Phase := 1; //Entry done
    else if pre(SequenceStep2Phase) == 1 then
        //Step activity code
        SequenceStep2Phase := 2; //Step done
    else if pre(SequenceStep2Phase) == 2 and true
        //Exit activity code
        SequenceStep2Phase := 3; //Exit done
        SequenceStep2 := false;
        SequenceFinalStep := true;
    end if;
else if SequenceFinalStep then
    SequenceFinalStep := false;
    Sequence := false;
    //A next Step/Sequence would be enabled here
    //Allocations releasing code here
end if;
```

Figure 5: An automation sequence diagram with a corresponding (Modelica) algorithm section.

Steps, Allocations, sub-Sequences, and pseudo Steps are handled with conditional (if-else if) code blocks that can be all entered only once. This is necessary because Modelica models are executed cyclically. In a cycle the execution must continue from the phase to which the execution ended in the previous cycle. For example, arriving to the allocation phase in the example is enabled in the initialization phase and disabled in the allocation phase, which in turn enables the next phase.

Entry, Step, and Exit Activities are executed only once so that when arriving to a Step, the Entry Activity is executed first in addition to changing the phase value to 1. Next, Step Activity is executed and the phase value set to 2. The execution of the Exit Activity and setting the phase variable to 3 waits until the transition condition (if any) to next Step in the Sequence is satisfied so that the transition can occur immediately after performing the Exit Activities.

If the Step in question does not contain Entry, Step, or Exit Activities, the corresponding algorithmic code only changes the value of the phase variable.

Allocations are assumed to be next to initial Steps in Sequences. They are intended to model allocations of devices that have IDs corresponding to the ID variables of Allocations. For Allocations, the algorithmic code increases (by one) the variables of the record that correspond to the allocated IDs. At the end of Sequences, allocations are relieved by decreasing the values of the variables by one. In the simulations, the Allocations thus do not force execution to wait but only warn about double allocations, which are indicated by the values of the variables becoming greater than 1. Such problems can then be inspected by developers.

Fork-to-join regions are in the approach handled by creating variables for each branch in the region. The branches may execute independently of each other but for a transition to exit a join Step, all branches must have reached the join. This condition is used as an exit guard for the join, in addition to possible transition conditions related to the transition exiting it.

In the approach, the Modelica code structures resemble the structures in [31] that are used for Modelica simulating state machines. The most notable differences are as follows. Steps or sub-Sequences that are next to another sub-Sequence are not enabled until the sub-Sequence reaches its final Step. This prevents a transition in a higher-level sequence to fire before the final Step is reached. The phases of Steps, that is, whether the Entry, Step, and Exit Activities have been executed, are recorded with Integer variables. The transition conditions to exit Steps are used inside the Steps as guards for shifting to the Exit Activities and enabling the next Step/sub-Sequence. In Allocations that do not have activities the transition conditions are similarly used as conditions to enable the next Steps or sub-Sequences. In referenced sub-Sequences, the next Steps or sub-Sequences are enabled in the final Steps. Lastly, in case of a transition containing a timeout condition, a real valued time variable is created for the previous Step the value of which is set to equal the completion time of the previous Step, pseudo Step, or sub-Sequence. The time variables can be used in the transition conditions as illustrated in Figure 5.

Constraints and Assumptions

In development of the modeling and simulation approach, a decision was made that Sequences must always be owned by AFs. In this way, the variables and algorithmic code corresponding to a Sequence can be created for a Modelica class corresponding to the AF that owns the Sequence. In the approach a Sequence thus describes the sequential behavior of the AF that owns the Sequence. A Step being executed in a Sequence is a Step of the AF. For control or other signals to be forwarded to other AFs, the AF must be connected to them with use of ports in the interfaces of the AFs. The execution of a single Sequence is thus centralized in an AF. However, an AF may contain several Sequences. On the other hand, a control application model may contain several AFs that define Sequences so that at runtime there would be several Sequences executing concurrently and independently of each other.

The properties that are created for implementing the sequential behavior (see previous section) become the properties of the (ModelicaML) class that is created to correspond to the AF. The properties are necessary for implementing the dynamic behavior, by controlling the execution of algorithmic statements. During simulation, however, they also indicate the execution of the Sequence. For example, the Boolean valued properties created to correspond to Sequences (and its possible sub-Sequences) have value true only when the execution is in the sub-Sequence in question. This feature has been used, for example, in Figure 9 in which the upper plot presents the sub-Sequences of a pulp batch processing Sequence.

There are also restrictions related to the use of Sequences in the approach. Currently, for the simulation transformation to work properly, fork-to-join regions must be balanced so that branches exiting a fork meet each other in one join. On the other hand, the transformation does not support loops within Sequences so that a Step could be entered more than once in a Sequence. It is also assumed that a Sequence always contains an initial Step and at least one final Step. Whether the restrictions related to initial and final Steps hold is checked before performing the transformation.

Implementation of the Approach

The transformation for simulating Sequences was implemented by extending the previous version of the transformation [1]. In addition to Control Structure and logic diagrams that were supported by the previous version, the transformation processes Sequences contained by AFs to properties and algorithm sections of Modelica classes. The core of the transformation was written with the QVT (Query/View/Transformation) operational mappings language [29] and the executable Java code generated with the SmartQVT tool. The generated Java transformation class was complemented by extending it with a manually written class. It takes care of, for example, the creation of tagged values of stereotypes and other tasks that are hard to express with QVT languages. To enable launching the transformation with the graphical user interface of the supporting UML AP tool [28], the transformation was packaged to an Eclipse plugin. The plugin architecture and integration to the tool was implemented as outlined in [32].

ILLUSTRATIVE EXAMPLE

The example system to be used in this paper is a laboratory scale pulp processing plant the piping and instrumentation (P&I) diagram of which is in Figure 6. The plant includes 3 storage tanks, a boiler, 2 pumps, 2 control valves, 13 solenoid valves, and piping that enable pumping fluid from any tank to the boiler and via boiler back to any of the tanks. The tanks contain instrumentation to measure liquid levels in them, temperature in tank T100, and pressure in boiler T300. The process is used to simulate batch processing of pulp which is located in the boiler and processed with process substances (impregnation liquor, black liquor, and white liquor) according to timing, pressure, and temperature constraints. In specific phases of the processing sequence, feedback control is required to control the temperature of the white liquor (in tank T100) and pressure in the boiler (T300).

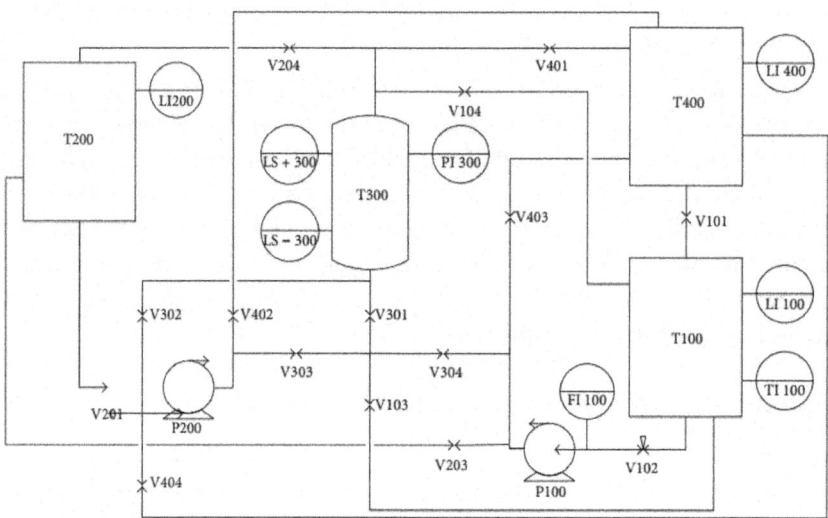

Figure 6: P&I diagram of the pulp production process.

To enable the simulation of the process with a modelled control solution, the process was modeled with ModelicaML. This included defining simulation classes for the physical parts of the process including tanks, boiler, pumps, solenoid valves, control valves, pipes, and pipe crossings with 3 and 4 inlets. Tanks keep record on liquid levels and temperatures inside them. For temperature equations, ideal mixing of fluids is assumed. The liquid flows in pipes and in control valves that are proportional to constants measured from the process and to square roots of the pressure differences between the ends of the pipes/valves. Pumps increase the pressure in their output sides and solenoid valves stop the liquid flows regardless of the pressure differences.

The simulatable ModelicaML model was then defined by creating instances of the classes and connecting them together according to the connections in the physical process. This was done with a structured class diagram. A small part of the diagram, related to the surroundings of the tank T400, is presented in Figure 7.

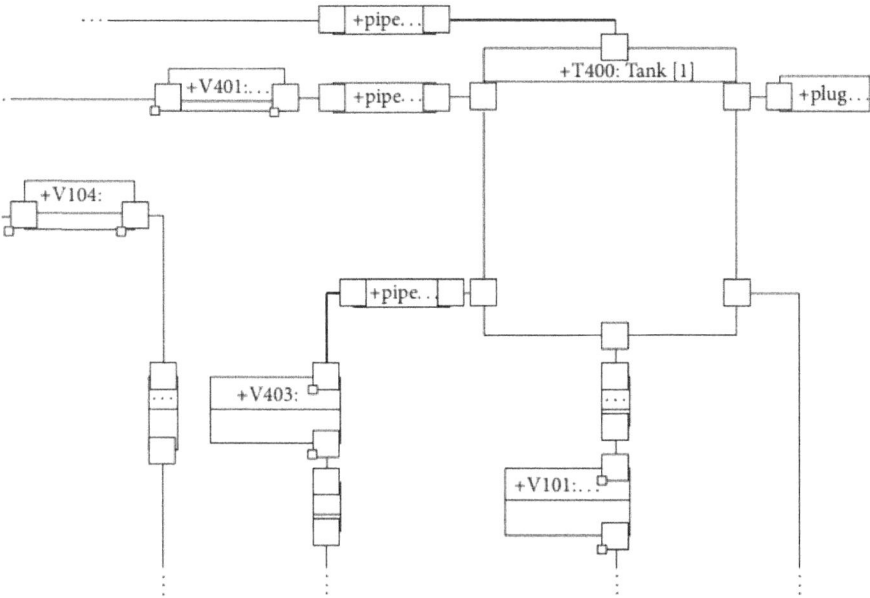

Figure 7: A part of the ModelicaML plant model.

The control solution for the batch process is illustrated with Figures 8 and 3. Figure8 presents a (UMLAP) Control Structure Diagram of the control solution. It contains binary and analogue valued input and output AFs for interfacing with the sensors and actuators of the process. The Sequence is implemented within the MPPSequence AF that controls some actuators directly and uses controllers for controlling T300 pressure (by throttling valve V104) and T100 temperature with heater E100. To illustrate how logic diagrams and ASDs are used to define behavior of AFs, the figure has been complemented with the MPP Sequence and a logic diagram definition of the temperature controller.

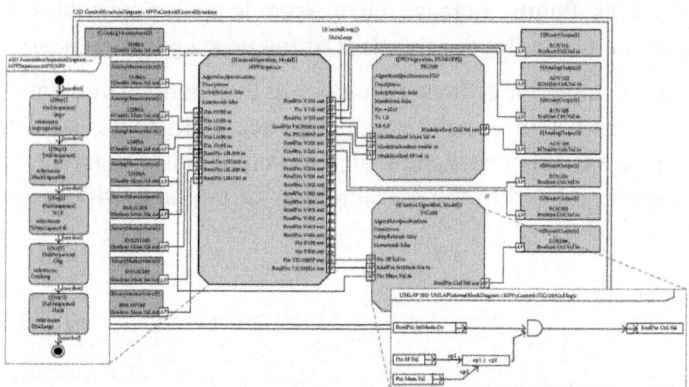

Figure 8: Modeling of automation sequences integrates to previous work with control structure and logic diagrams.

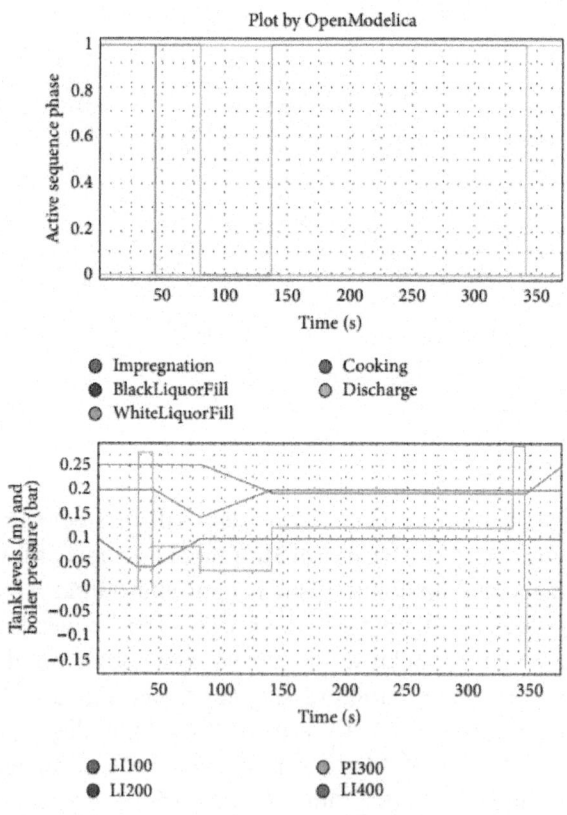

Figure 9: Simulation results plotting active phases of the sequence, levels in tank T100 (LI100), T200 (LI200), and T400 (LI400) as well as pressure in boiler T300 (PI300).

The other illustrating figure (Figure 3) was used as an example earlier and illustrates the MPPSequence and one of the sub-Sequences of it, WhiteLiquorFill. MPPSequence consists of 5 main phases: Impregnation, BlackLiquorFill, WhiteLiquorFill, Cooking, and Discharge. During the phases, the boiler is filled with impregnation liquor and pressurized, filled with black liquor that replaces the impregnation liquor (BlackLiquorFill), filled with white liquor that replaces the black liquor (WhiteLiquorFill), heated to cooking temperature and pressurized (Cooking), and finally drained back to white liquor tank T400 (Discharge). WhiteLiquorFill, on the other hand, opens valves V301, V401, V102, and V304, and pumps liquor until the level in tank T400 exceeds 0.2 (m).

In order to obtain simulation results of a closed-loop system, the developed transformation was used to transform and connect the control system model to the plant model. Practically this included selecting the simulator export functionality of the tool and the (target) ModelicaML plant model file. After performing the transformation, the model was simulated with OpenModelica [26] tools. The ModelicaML model was first transformed to Modelica code and then loaded to the simulator environment. Initial values for the plant, for example, levels and temperatures in the tanks, were defined in the process model. For different simulations they could have been changed at this point too.

A plot illustrating the results from simulating the Sequence is shown in Figure 9. The main phases are plotted in the upper part of the figure, a value being one indicating execution of the phase in question. The lower part of the figure plots the levels of liquor in tanks T100 (LI100), T200 (LI200), and T400 (LI400) and pressure in tank T300 (PI300). According to the results, the control solution including the Sequence works as intended. Processing liquors are used in the correct order and the boiler pressurize, during black liquor fill and cooking phases.

The values shown in the figure were selected for plotting after performing the simulation. The simulator keeps record on all variables related to a simulation. Any other set of variables related to an aspect in the process or in the control solution, for example, functioning of a controller, could have been selected for plotting as well.

DISCUSSION

This paper has addressed the issue of simulating sequential control activities within MDD of control applications. The approach integrates to the previous work of the authors and enables the use of Automation Sequence Diagrams (ADSs) of UML AP to define sequential behavior of Automation Functions for simulation purposes. The transformation to simulatable ModelicaML form was implemented with open source modeling and model transformation (QVT) tools on the Eclipse platform. The ASD diagram type that is in the approach used for modeling sequential control has been extended from UML state machine diagrams. However, because of significant differences in execution semantics of state machines, it was not possible to rely on existing work [31] related to simulating them in Modelica form.

The benefits from using Modelica (ML) as the (target) simulation language of the approach included the ability to use standard model transformation techniques. Modelica is also an object-oriented simulation language, which was taken the advantage of mainly in development of the plant simulation model. From the point of view of simulating the control application, however, object-oriented features were not used. As a consequence, it is expected that the presented approach could be used also with other simulation languages that can be accessed with model transformations, for example, Simulink. An approach to execute Sequences without equation based, acausal execution semantics of Modelica could also be similar to the one presented in this paper. Algorithmic constructs were used also in case of Modelica instead of equations that apply all the time.

The novelty of the simulation approach is in the ability to simulate control application models at design time, before IEC language [17, 18] implementations of the applications. Closed-loop MiL simulations are created with model transformations so that a genuine simulation language (Modelica) is used for simulating both plant and control application models. Other MDD approaches in the domain (in which simulations have been supported) have utilized IEC 61499 as a simulation (in addition to implementation) language [15, 16, 20] or relied on the use of cosimulation [14, 19]. On the other hand, sequential control as a special aspect of control systems has been addressed only in [13] but not with respect to simulations. With the work presented in this paper and [1], the simulation approach covers all the common aspects of basic control systems including binary and feedback control, sequential control, and interlocks.

An issue that is not yet addressed in the approach [1] is delays in control systems hardware, for example, networks in distributed control systems. However, the objective of the approach is to enable simulations early, already before, for example, finishing control system hardware design. On the other hand, effects of delays and, for example, random noise in instrumentation can be included in the models, in simulation classes of sensors and actuators of the process. It is also assumed that, for example, delays in typical control system hardware are less significant than those in instrumentation.

Support for model-based control software development is also part of some commercial products. For example, B&R (Automation Studio) [33] and Beckhoff (TwinCAT 3) [34] support the development of control applications in MATLAB/Simulink environment and generating executable (PLC) code based on the models. As a difference to such products, the work presented in this paper intends to support simulations in an MDD approach in which all models are not simulatable. Instead, models are developed gradually from requirements towards executable applications using model transformations for shifting between models and, for example, importing source information to models. In addition, the models cover special needs such as traceability between requirements and design artifacts that are becoming more and more important in the domain.

To illustrate the simulation approach, it was applied to simulation of a controlled pulp batch production process. For the case study, the pulp production process was modelled with Modelica. Flow, pressure, and temperature equations for all the plant components in the model led to the total number of equations for the closed-loop system to be approximately 1400. As such, the closed-loop system was the largest that has been utilized in the simulation experiments of the approach so far. It also demonstrates the scalability of the approach for practical, nontrivial simulation needs.

CONCLUSIONS

MDD techniques are under active research in the application domain of industrial control systems. However, despite the research activities, and the tradition of using simulations, simulations have not yet been sufficiently integrated to MDD in the domain.

In MDD, it is possible to utilize model transformations for obtaining simulation models already before programmed implementations of the applications. This possibility should be taken advantage of. Control applications models should be evaluated in a timely manner and in closed-loops with the models of the processes to be controlled. In order to relieve control application developers from the task of coupling simulation engines, the simulations should follow the model-in-the-loop approach using a single simulation engine.

The presented approach complements the simulation approach of the authors with the possibility to simulate sequential control activities in conjunction to feedback and binary control as well as interlocks. The new work has been targeted for the sequences of process and batch industry. However, control sequences can be beneficial also in simulations of other kinds of processes. For example, in a previous simulation experiment [1], the set-point trajectories to evaluate a control system in different conditions needed to be defined manually. With the work presented, the set-point trajectories can be included in Sequences of the models.

According to our experiences, the simulation approach is useful in revealing defects in control algorithms, structures, and tunings. The simulations can be performed already at design time and so that decisions made in a development phase can be evaluated before they affect decisions in later phases. By creating simulation models with automated model transformations, simulations can be used as a continuous, design-time quality assurance method. This can be done without causing excessive additional workload to developers.

It is also expected that the task of developing models of the processes to be controlled with Modelica becomes easier and more attractive for industry in near future. This is due to improvements in libraries of simulation classes, the Modelica standard library, from which it is possible to compose plant models. It is also a clear benefit of Modelica that it includes support for standard and user/company specific libraries. Modelica is already supported by both commercial and open source tools that can be used by both industry and academy.

CONFLICT OF INTERESTS

The authors declare that there is no conflict of interests regarding the publication of this paper.

REFERENCES

1.   T. Vepsäläinen and S. Kuikka, "Simulation-based development of safety related interlocks," in Simulation and Modeling Methodologies, Technologies and Applications, pp. 165–182, Springer, 2013.

2.   D. Hästbacka, T. Vepsäläinen, and S. Kuikka, "Model-driven development of industrial process control applications," Journal of Systems and Software, vol. 84, pp. 1100–1113, 2011.

3.   T. Ritala and S. Kuikka, "UML automation profile: enhancing the efficiency of software development in the automation industry," in Proceedings of the 5th IEEE International Conference on Industrial Informatics (INDIN ‹07), pp. 885–890, June 2007.

4.   H. Shokry and M. Hinchey, "Model-based verification of embedded software," Computer, vol. 42, no. 4, pp. 53–59, 2009.

5.   Plummer, "Model-in-the-loop testing," Proceedings of the Institution of Mechanical Engineers I, vol. 220, pp. 183–199, 2006.

6.   H. Chae, X. Jin, S. Lee, and J. Cho, "TEST: testing environment for embedded systems based on TTCN-3 in SILS," Communications in

Computer and Information Science, vol. 59, pp. 204–212, 2009.

7. M. Short and M. J. Pont, "Assessment of high-integrity embedded automotive control systems using hardware in the loop simulation," Journal of Systems and Software, vol. 81, no. 7, pp. 1163–1183, 2008.

8. M. Schlager, R. Obermaisser, and W. Elmenreich, "A framework for hardware-in-the-loop testing of an integrated architecture," in Software Technologies for Embedded and Ubiquitous Systems, pp. 159–170, Springer, 2007.

9. G. Stoeppler, T. Menzel, and S. Douglas, "Hardware-in-the-loop simulation of machine tools and manufacturing systems," IEE Computing and Control Engineering, vol. 16, no. 1, pp. 10–15, 2005.

10. J. A. Carrasco and S. Dormido, "Analysis of the use of industrial control systems in simulators: state of the art and basic guidelines," ISA Transactions, vol. 45, no. 2, pp. 295–312, 2006.

11. MODELISAR Consortium, "Functional Mock-up Interface for Co-simulation," Version 1.0., 2010.

12. G. Hemingway, H. Neema, H. Nine, J. Sztipanovits, and G. Karsai, "Rapid synthesis of high-level architecture-based heterogeneous simulation: a model-based integration approach," Simulation, vol. 88, no. 2, pp. 217–232, 2012.

13. T. Lukman, G. Godena, J. Gray, M. Heričko, and S. Strmčnik, "Model-driven engineering of process control software-beyond device-centric abstractions,"Control Engineering Practice, vol. 21, no. 8, pp. 1078–1096, 2013.

14. H. Thompson, D. Ramos-Hernandez, J. Fu, L. Jiang, J. Nu, and D. Dobinson, "The FLEXICON co-simulation tools applied to a marine application,"Proceedings of the Institution of Mechanical Engineers M: Journal of Engineering for the Maritime Environment, vol. 222, pp. 81–94, 2008.

15. V. Vyatkin, H. Hanisch, C. Pang, and C. Yang, "Closed-loop modeling in future automation system engineering and validation," IEEE Transactions on Systems, Man and Cybernetics Part C: Applications and Reviews, vol. 39, no. 1, pp. 17–28, 2009.

16. Hegny, M. Wenger, and A. Zoitl, "IEC 61499 based simulation framework for model-driven production systems development," in Proceedings of the 15th IEEE International Conference on Emerging Technologies and Factory Automation (ETFA ‹10), September 2010.

17. International Electrotechnical Commission, IEC 61499-1: Function

Blocks-Part 1: Architecture, International Standard, Geneva, Switzerland, 1st edition, 2012.

18. International Electrotechnical Commission, IEC 61131-3: Programmable Controllers part 3, Programming Languages, IEC Publication, 2013.

19. L. Ferrarini and A. Dedè, "A model-based approach for mixed Hardware in the Loop simulation of manufacturing systems," in Proceedings of the 10th IFAC Workshop on Intelligent Manufacturing Systems (IMS ‹10), pp. 36–41, July 2010.

20. C. Yang and V. Vyatkin, "Transformation of Simulink models to IEC 61499 Function Blocks for verification of distributed control systems," Control Engineering Practice, vol. 20, no. 12, pp. 1259–1269, 2012.

21. Provost, J. Roussel, and J. Faure, "Translating Grafcet specifications into Mealy machines for conformance test purposes," Control Engineering Practice, vol. 19, no. 9, pp. 947–957, 2011.

22. R. David, "Grafcet: a powerful tool for specification of logic controllers," IEEE Transactions on Control Systems Technology, vol. 3, no. 3, pp. 253–268, 1995.

23. Hellgren, M. Fabian, and B. Lennartson, "On the execution of sequential function charts," Control Engineering Practice, vol. 13, no. 10, pp. 1283–1293, 2005.

24. P. Fritzson and V. Engelson, "Modelica—a unified object-oriented language for system modeling and simulation," in Proceedings of the Object-Oriented Programming Conference (ECOOP '98), pp. 67–90, Springer, 1998.

25. W. Schamai, Modelica Modeling Language (ModelicaML): A UML Profile for Modelica, Linköping University Electronic Press, 2009.

26. OpenModelica, https://www.openmodelica.org/.

27. Dassault Systemes, Dymola, http://www.3ds.com/products-services/catia/capabilities/systems-engineering/modelica-systems-simulation/dymola.

28. T. Vepsäläinen, D. Hästbacka, and S. Kuikka,, "Tool support for the UML automation profile—for domain-specific software development in manufacturing," in Proceedings of the 3rd International Conference on Software Engineering Advances (ICSEA ‹08), pp. 43–50, 2008.

29. OMG, "Meta Object Facility (MOF) 2.0 Query/View/Transformation Specification (QVT), Version 1.0," Object Management Group, 2008.

30. T. Vepsäläinen and S. Kuikka, "Benefit from simulating early in MDE of industrial control," in Proceedings of the IEEE 18th Conference on

Emerging Technologies & Factory Automation (ETFA ‹13), pp. 1–8, 2013.

31. W. Schamai, U. Pohlmann, P. Fritzson, C. J. Paredis, P. Helle, and C. Strobel, "Execution of UML State machines using modelica," in Proceedings of the 3rd International Workshop on Equation-Based Object-Oriented Modeling Languages and Tools (EOOLT ‹10), pp. 1–10, 2010.

32. T. Vepsäläinen, D. Hästbacka, and S. Kuikka, "A model-driven tool environment for automation and control application development-transformation assisted, extendable approach," in Proceedings of 11th Symposium on Programming Languages and Software Tools and 7th Nordic Workshop on Model Driven Software Engineering, pp. 315–329, 2009.

33. B&R Automation Studio, http://www.br-automation.com/en/products/software/automation-studio/.

34. Beckhoff TwinCAT 3, http://www.beckhoff.fi/english.asp?twincat/default.htm.

Chapter 7

A REVIEW OF ACTIVE YAW CONTROL SYSTEM FOR VEHICLE HANDLING AND STABILITY ENHANCEMENT

M. K. Aripin,[1] Yahaya Md Sam,[2] Kumeresan A. Danapalasingam,[2] Kemao Peng,[3] N. Hamzah,[4] and M. F. Ismail[5]

[1]Control, Instrumentation & Automation Department, Faculty of Electrical Engineering, Universiti Teknikal Malaysia Melaka, 76100 Durian Tunggal, Melaka, Malaysia

[2]Department of Control & Mechatronics, Faculty of Electrical Engineering, Universiti Teknologi Malaysia, 81310 UTM Johor Bahru, Johor, Malaysia

[3]Temasek Laboratories, National University of Singapore 5A Engineering Drive 1, Singapore 117411

[4]Faculty of Electrical Engineering, UiTM Pulau Pinang, 13500 Permatang Pauh, Pulau Pinang, Malaysia

[5]Industrial Automation Section, Universiti Kuala Lumpur Malaysia France Institute, Section 14, Jalan Teras Jernang, 43650 Bandar Baru Bangi, Selangor, Malaysia

ABSTRACT

Yaw stability control system plays a significant role in vehicle lateral dynamics in order to improve the vehicle handling and stability performances. However, not many researches have been focused on the transient performances improvement of vehicle yaw rate and sideslip tracking control. This paper reviews the vital elements for control system design of an active yaw stability control system; the vehicle dynamic models, control objectives, active chassis control, and control strategies with the focus on identifying suitable criteria for improved transient performances. Each element is discussed and compared in terms of their underlying theory, strengths, weaknesses, and applicability. Based on this, we conclude that the sliding mode control with nonlinear sliding surface based on composite nonlinear feedback is a potential control strategy for improving the transient performances of yaw rate and sideslip tracking control.

INTRODUCTION

In vehicle dynamic control of road-vehicle, controlling the lateral dynamic motion is very important where it will determine the stability of the vehicle. One of the prominent approaches that are reported in the literature for lateral dynamics control is a yaw stability control system. In order to design an effective control system, it is essential to determine an appropriate element of yaw stability control system. In this paper, the elements of yaw stability control system, that is, vehicle dynamic models, control objectives, active chassis control, and its control strategies as depicted in Figure 1, are extensively reviewed.

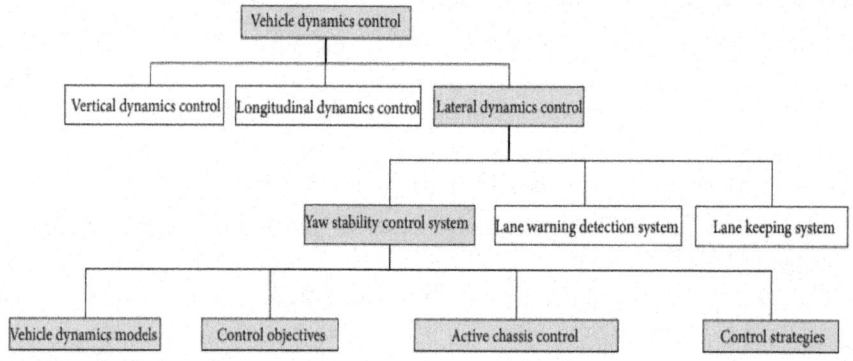

Figure 1: Yaw stability control system for vehicle lateral dynamic.

The linear and nonlinear vehicle models that described the behaviour of lateral dynamic are explained for controller design and evaluation purpose. To achieve the control objectives, it is essential to control the variables of yaw rate and sideslip angle in order to ensure the vehicle stable. It is required that the actual yaw rate and sideslip angle have fast responses and good tracking capability in following the desired responses. During critical driving condition or manoeuvre, inappropriate commands by the driver to control the steering and braking can cause the vehicle to become unstable and lead to an accident. Therefore, an active control for yaw stability control system is essential to assist the driver to keep the vehicle stable on the desired path. By implementing an active chassis control of steering or braking or integration of both systems, the active yaw control system can be realized.

In real driving condition, the lateral dynamics of vehicle are incorporated with uncertainties such as different road surface condition, varying vehicle parameters, and crosswind disturbance. In yaw stability control system, these perturbations could influence the yaw rate and sideslip tracking control performances. From the control system point of view, the transient performances of tracking control are essential. However, from the reviewed control strategies in the literature, the controllers are not designed to cater this matter. Therefore, an appropriate robust control strategy should be proposed to improve the transient performances of the yaw rate and sideslip tracking control in the presence of uncertainties and disturbances. As a finding from the reviews, this paper briefly discussed a possible high performance robust tracking control strategy that can be implemented for yaw stability control system.

The review begins with vehicle dynamics models in Section 2. The yaw stability control objectives are discussed in Section 3 and followed by active chassis control for in Section 4. Yaw stability control strategies and problems are reviewed in Sections 5 and 6, respectively. In Section 7, a high performance robust tracking controller using sliding mode control and composite nonlinear feedback is discussed. The controller evaluation is discussed in Section 8 and ended with conclusion in Section 9.

VEHICLE DYNAMICS MODELS

In order to examine, analyse, and design the controller for yaw stability control system, vehicle dynamics models are essential where the mathematical modelling of vehicle dynamic motion is obtained based on Newton's 2nd law that describes the forces and moments acting on the vehicle body and tires. In general, there are two categories of vehicle dynamic model, that is, nonlinear vehicle model and linearized vehicle model as depicted in Figure 2. The following subsections will discuss the nonlinear vehicle model for simulation and linearized vehicle model for controller design purpose.

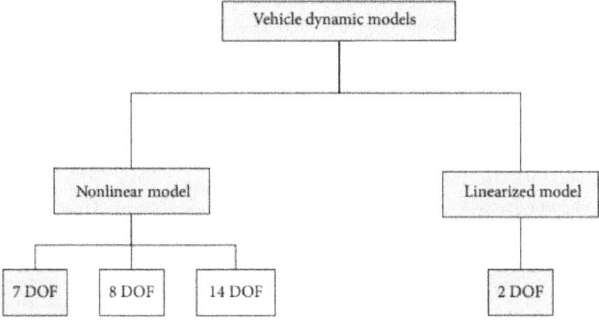

Figure 2: Vehicle dynamic models.

Vehicle Model for Simulation

The nonlinear vehicle model is regularly used to represent and simulate the actual vehicle for controller evaluation and validation. In recent years, researches in [1–5] have utilized nonlinear vehicle model for vehicle handling and stability improvement studies. Figure 3 shows the typical nonlinear vehicle model in cornering manoeuvre gravity (CG) l_f and l_r, respectively.

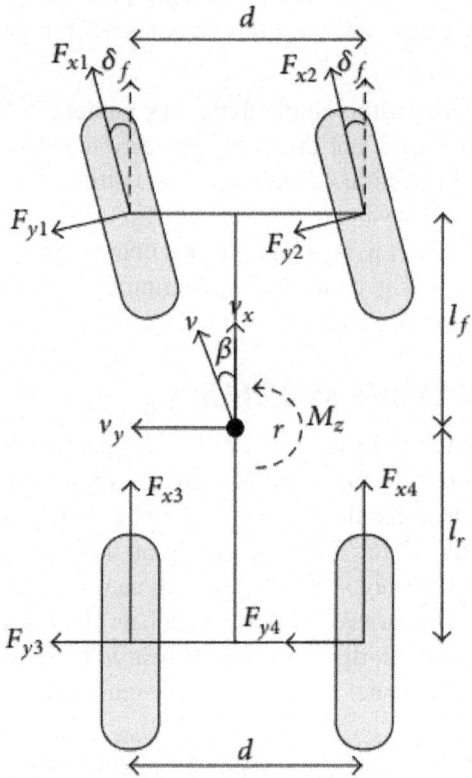

Figure 3: Nonlinear vehicle model [6].

The input of this model is front wheel steer angle δf while the output variables to be controlled are vehicle sideslip β and yaw rate r. The vehicle parameters are vehicle width track d, distance from front, and rear axle to centre of

The vehicle forward velocity of centre of gravity (CG) is V, lateral velocity is V_y, and longitudinal velocity is V_x. Other important vehicle parameters are vehicle mass m, moment of inertia Iz, and front/rear tire cornering stiffness C_f/ C_r. The wheels are numbered as subscript number with (1) for front-left, (2) for front-right, (3) for rear-left, and (4) for rear-right. Longitudinal tire force, F_{xi}, depends directly on tire slip ratio, λ_i while lateral tire force, F_{yi}, depends directly on tire sideslip angle, α_i. For smaller slip angle and slip ratio, lateral tire force is described as a linear function of the tire cornering stiffness and tire sideslip angle while longitudinal tire force is described as a linear function of the braking stiffness and the tire slip ratio. For larger slip angle and slip ratio, longitudinal and lateral tire forces exhibit a nonlinear characteristics. Vehicle dynamic motion with nonlinear tire forces represents a nonlinear system. The nonlinear lateral and longitudinal tire forces can be described using prominent Pacejka tire model as implemented in [1, 4, 7] or Dugoff tire model as utilized in [8–10], while studies in [11] used both tire models

The nonlinear vehicle model could have different number of degree-of-freedom (DOF) where it represents the dynamics motions and complexity of vehicle models. As utilized in [2, 12–14], the 7 DOF vehicle model represents the dynamic motions of vehicle body, that is, longitudinal, lateral, yaw, and four wheels. The dynamic equations for the longitudinal, lateral, and yaw motions of the vehicle body are described as follows.

Longitudinal Motion. One has the following:

$$ma_x = m\left(\dot{v}_x - rv_y\right)$$
$$= \left(F_{x1} + F_{x2}\right)\cos\delta_f + F_{x3} + F_{x4} - \left(F_{y1} + F_{y2}\right)\sin\delta_f. \tag{1}$$

Lateral Motion. One has the following

$$ma_y = m\left(\dot{v}_y + rv_x\right)$$
$$= \left(F_{x1} + F_{x2}\right)\sin\delta_f + \left(F_{y1} + F_{y2}\right)\cos\delta_f + F_{y3} + F_{y4}. \tag{2}$$

Yaw Motion. One has the following:

$$I_z\dot{r} = l_f\left(F_{y1}\cos\delta_f + F_{y2}\cos\delta_f + F_{x1}\sin\delta_f + F_{x2}\sin\delta_f\right)$$
$$- l_r\left(F_{y3} + F_{y4}\right) + M_z, \tag{3}$$

where M_z is yaw moment that must be taken into account, that is; $M_z > 0$ if the tires tends to turn at z-axis. In (2), the lateral acceleration ay can be expressed in terms of vehicle forward speed V, yaw rate r, and sideslip β as follows:

$$a_y = \dot{v}_y + rv_x = v\left(r + \dot{\beta}\right). \tag{4}$$

Therefore, the output variable of sideslip β of two-track model can be obtained as follows:

$$\dot{\beta} = \frac{1}{mv} \left[\cos\beta \left(\cos\delta_f \left(F_{x1} + F_{x2} \right) - \sin\delta_f \left(F_{y1} + F_{y2} \right) \right) \right.$$

$$\left. - \sin\beta \left(\sin\delta_f \left(F_{x1} + F_{x2} \right) - \sin\delta_f \left(F_{y1} + F_{y2} \right) \right) \right]$$

$$- r. \tag{5}$$

while the output variable of yaw rate r can be determined from (3) and obtained as follows:

$$\dot{r} = \frac{1}{I_z} \left[l_f \left(F_{y1} \cos\delta_f + F_{y2} \cos\delta_f + F_{x1} \sin\delta_f + F_{x2} \sin\delta_f \right) \right.$$

$$\left. - l_r \left(F_{y3} + F_{y4} \right) + M_z \right]. \tag{6}$$

In vehicle dynamic studies, each wheel represents 1 DOF. Thus, there are 4 DOF for road-vehicle with 4 wheels. The dynamic motion for each wheel is described as follows:

$$I_{wi}\dot{\omega}_i = -R_{wi}F_{xi} + T_{ei} - T_{bi}, \tag{7}$$

where ω is wheel angular acceleration, n, R_w is wheel radius, Iw is wheel inertia, T_{bi} is braking torque, and T_{ei} is driving torque

Another nonlinear vehicle model used in the previous research is 8 DOF vehicle model that is extensively used in [4, 5, 9–11, 15–18]. For more accurate simulation and validation, the 14 DOF vehicle model is used in [1, 19, 20]. The comparison between the number of DOF of nonlinear vehicle models that have been discussed above can be summarized and compared in Table 1.

Another nonlinear vehicle model used for simulation uses a multi-degree-of-freedom vehicle model based on commercial vehicle dynamics software, that is, CarSim, as implemented in [21–26]. By using this software based vehicle model, the dynamic behaviour of vehicle is more precise similar to a real vehicle. However, for yaw rate and sideslip tracking control in yaw stability control system, the 7 DOF nonlinear vehicle model as discussed in the above equations and shown in Table 1 is adequate for simulation and evaluation of the design controller. 2.2. Vehicle Model for Controller Design. In vehicle dynamic studies, the classical bicycle model as shown in Figure 4 is prominently used for yaw stability control analysis and controller design as utilized in [1, 3, 8, 26–30]. This model is linearized from the nonlinear vehicle model based on the following assumptions.

Table 1: Number DOF of nonlinear vehicle models.

Number of DOF	Dynamic motions	Output variable
7 DOF	(i) Longitudinal (ii) Lateral (iii) Vertical (iv) Rotational of 4 wheels	Yaw rate & Sideslip
8 DOF	(i) Longitudinal (ii) Lateral (iii) Vertical (iv) Roll (v) Rotational of 4 wheels	Yaw rate, roll rate, and sideslip
14 DOF	(i) Longitudinal (ii) Lateral (iii) Vertical (iv) Roll (v) Pitch (vi) Bounce (vii) Rotational of 4 wheels (viii) Vertical oscillations of 4 wheels	Yaw rate, roll rate, pitch rate, and sideslip

(i)   Tires forces operate in the linear region.

(ii)   The vehicle moves on plane surface/flat road (planar motion).

(iii)   Left and right wheels at the front and rear axle are lumped in single wheel at the centre line of the vehicle.

(iv)   Constant vehicle speed i.e. the longitudinal acceleration equal to zero ($ax=0$)

(v)   Steering angle and sideslip angle are assumed small (≈ 0).

(vi)   No braking is applied at all wheels.

(vii)   Centre of gravity (CG) is not shifted as vehicle mass is changing.

(viii) 2 front wheels have the same steering angle.

(ix)   Desired vehicle sideslip is assumed to be zero in steady state.

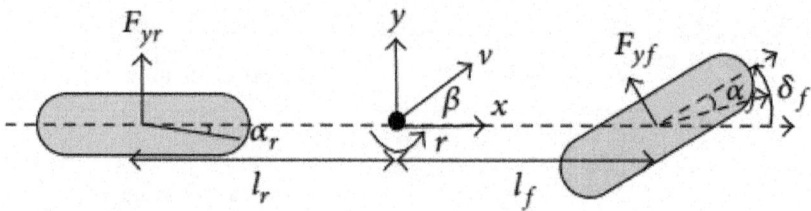

Figure 4: Bicycle model [31].

In the simplest form of planar motion, this model consists of 2 DOF for lateral and yaw motions as describe in the following equations. Lateral Motion. One has the following:

$$mv\left(\dot{\beta} + r\right) = \left(F_{yf} + F_{yr}\right) - r.$$ (8)

Yaw Motion. One has the following:

$$I_z \dot{r} = l_f \cdot F_{yf} - l_r \cdot F_{yr}.$$ (9)

In this model, the front and rear lateral tire forces F_{yf} and F_{yr}, respectively, exhibit linear characteristics and described as a linear function of the front and rear cornering stiffness, C_f and C_r, as follows:

$$F_{yf} = C_f \alpha_f,$$
$$F_{yr} = C_r \alpha_r,$$ (10)

where the front and rear tire sideslip angle, α_f, and α_r for linear tire forces are given in the following equations:

$$\alpha_f = \delta_f - \beta - \frac{l_f r}{v},$$

$$\alpha_r = -\beta + \frac{l_r r}{v}.$$ (11)

By rearranging and simplifying (8)–(11), the differential equations of sideslip and yaw rate variables can be simplified as a linear state space model as follows:

$$\dot{x} = Ax + Bu,$$

$$\begin{bmatrix} \dot{\beta} \\ \dot{r} \end{bmatrix} = \begin{bmatrix} a_{11} & a_{12} \\ a_{21} & a_{22} \end{bmatrix} \begin{bmatrix} \beta \\ r \end{bmatrix} + \begin{bmatrix} b_1 \\ b_2 \end{bmatrix} u,$$

$$\begin{bmatrix} \dot{\beta} \\ \dot{r} \end{bmatrix} = \begin{bmatrix} \dfrac{-C_f - C_r}{mv} & -1 + \dfrac{C_r l_r - C_f l_f}{mv^2} \\ \dfrac{C_r l_r - C_f l_f}{I_z} & \dfrac{-C_f l_f^2 - C_r l_r^2}{I_z v} \end{bmatrix} \begin{bmatrix} \beta \\ r \end{bmatrix}$$

$$+ \begin{bmatrix} \dfrac{C_f}{mv} \\ \dfrac{C_f l_f}{I_z} \end{bmatrix} \delta_f,$$

(12)

where β and r are state or output variables, C_f and C_r are front and rear tire cornering stiffness, respectively, m is vehicle

mass, I_z is moment of inertia, l_f and l_r are distance from front and rear axle to centre of gravity, respectively, V is vehicle speed, and front tire steer angle δ_f is the input u to the model. Notice that vehicle speed V is assumed always constant which means the vehicle is not involved with accelerating and braking. Hence, only lateral and yaw motions are analysed.

Besides that, the bicycle model is also regularly used as desired or reference model to generate the desired response of the yaw rate and sideslip angle based on steady state condition or approximated first order response. In designing the control strategy based on vehicle active chassis control, the linear state space model in (13) is essential.

Yaw Stability Control Objectives

A vehicle yaw rate r and sideslip angle β are significant variables in vehicle yaw stability control system. As stated in [32], control objectives of yaw stability control system may be classified into three categories, that is, yaw rate control, sideslip control, and combination of yaw rate and sideslip control as illustrated in Figure 5.

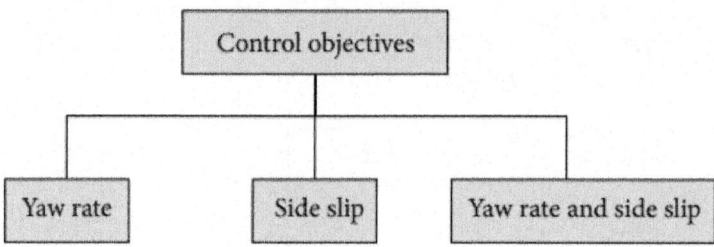

Figure 5: Yaw stability control objectives.

One of the control objectives of yaw stability control system is yaw rate, r. An ability to control the actual yaw rate close to desired response will improve the handling or manoeuvrability of the vehicle. The desired yaw rate which is generated by reference model should be tracked by the controller in order to improve the handling performance as mentioned in [2, 4, 13, 15, 18, 27, 33, 34]. In the steady state condition, the desired yaw rate response r_d can be obtained by using the following equation:

$$r_d = \frac{v}{\left(l_f + l_r\right) + k_{us} v^2} \cdot \delta_f,$$

(13)

where stability factor k_{us} is depending on the vehicle parameters and defined as follows:

$$k_{us} = \frac{m\left(l_r C_r - l_f C_f\right)}{\left(l_f + l_r\right) C_f C_r}.$$

(14)

Another control objective is the vehicle sideslip angle, β, that is, the deviation angle between the vehicle longitudinal axis and longitudinal axis and its motion direction. The control of sideslip angle close to steady state condition means controlling the lateral stability of the vehicle. For the steady state condition, the desired sideslip is always zero, that is, $\beta_d = 0$ as mentioned in [1, 6, 9, 11, 17, 26, 35]. Therefore, to improve the vehicle handling and stability performances, it

is essential to control both yaw rate and sideslip responses. In order to achieve these control objectives, the proposed controller must be able to perform the control task of the yaw rate and sideslip tracking control.

ACTIVE CHASSIS CONTROL

Steering and braking subsystems or actuator are part of the vehicle chassis. The active control of yaw stability control system can be realized through active chassis control, that is, direct yaw moment control or active steering control or integrated actives steering and direct yaw moment control as shown in Figure 6. In direct yaw moment control which can be implemented by active braking or active differential torque distribution, the required yaw moment is generated by the designed controller that controls the desired yaw rate and sideslip. In active steering control, the wheel steer angle that commanded by the driver is modified by adding corrective steer angle from the designed controller. This control strategy can be implemented either using active front steering (AFS) or active rear steering (ARS) or four-wheel active steering (4WAS) control. In order to control two variables of the yaw rate and sideslip effectively, two different control mechanisms are required. Thus, related research works on the integration of two vehicle chassis control, that is, integrated active steering and direct yaw moment control, have been extensively conducted recently. The review of direct yaw moment control, active steering control, and integrated active steering and direct yaw moment control are discussed in the following subsections.

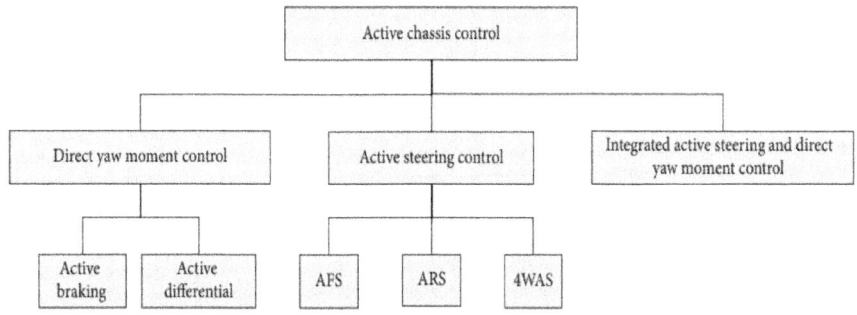

Figure 6: Active chassis control.

Direct Yaw Moment Control

Direct yaw moment control is one of the prominent methods for yaw stability control where extensive research works using this method have been conducted with different control strategies and algorithms as reported in [1, 3, 5, 8, 9, 15–18, 25, 26, 30, 36]. It is recognized as an effective method to enhance the vehicle lateral stability during critical driving manoeuvre by controlling the slip ratio of individual wheel. As illustrated in Figure 7, the required corrective yaw moment, ΔM_z which is generated by the transverse distribution of braking forces between the vehicle wheels is calculated by the designed controller based on the error between actual and desired vehicle model that have been discussed in Section 2. Another approach of direct yaw moment control is active distribution torque. By using an active differential device as established in [19, 20, 37, 38], the left-right driving torque is distributed by this device to generate the required corrective yaw moment, ΔM_z.

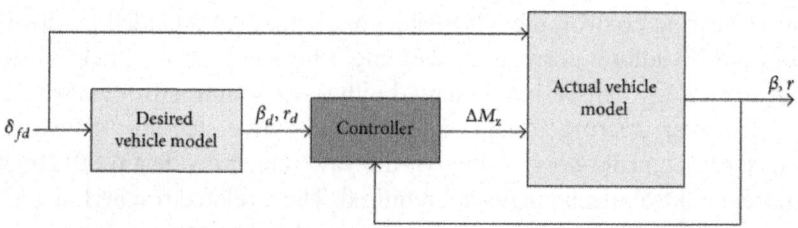

Figure 7: Direct yaw moment control [15].

As mentioned in Section 2, direct yaw moment control design is based on the linear state space model. As described in (15), M_z is considered as control input and front steer angle δ_{fd} is assumed as disturbance:

$$\begin{bmatrix} \dot{\beta} \\ \dot{r} \end{bmatrix} = \begin{bmatrix} a_{11} & a_{12} \\ a_{21} & a_{22} \end{bmatrix}\begin{bmatrix} \beta \\ r \end{bmatrix} + \begin{bmatrix} b_1 \\ b_2 \end{bmatrix}\delta_{fd} + \begin{bmatrix} b_3 \\ b_4 \end{bmatrix}M_z,$$

$$\begin{bmatrix} \dot{\beta} \\ \dot{r} \end{bmatrix} = \begin{bmatrix} \dfrac{-C_f - C_r}{mv} & -1 + \dfrac{C_r l_r - C_f l_f}{mv^2} \\ \dfrac{C_r l_r - C_f l_f}{I_z} & \dfrac{-C_f l_f^2 - C_r l_r^2}{I_z v} \end{bmatrix}\begin{bmatrix} \beta \\ r \end{bmatrix}$$

$$+ \begin{bmatrix} \dfrac{C_f}{mv} \\ \dfrac{C_f l_f}{I_z} \end{bmatrix}\delta_{fd} + \begin{bmatrix} 0 \\ \dfrac{1}{I_z} \end{bmatrix}M_z.$$

(15)

Although direct yaw moment control could enhance the vehicle stability for critical driving conditions, it may be less effective for emergency braking on split road surface. At high vehicle speed steady state cornering, direct yaw moment control could decrease the yaw rate and increase a burden to the driver. To overcome this disadvantage, active steering control is proposed.

Active Steering Control

Active steering control is another approach to improving the vehicle yaw stability, especially for steady state driving condition where the lateral tire force is operated in the linear region. Research works of active steering control have been continuously conducted in order to improve the handling and stability performances as reported in [7, 13, 39–42]. In general, active steering control can be divided into three categories, that is, active front steering (AFS) control, active rear steering (ARS) control, and four-wheel active steering (4WAS) control, as shown in Figure 6. As road-vehicle normally has front-wheel steering, AFS control becomes favourite approach among researchers as it can be combined with active braking and/or suspension control. In the AFS control diagram, as shown in Figure 8, the

front wheel steers angle is a sum of steer angle commanded by the driver δ_{fd} and a corrective steer angle δ_c generated by the controller. This corrective steer angle is computed based on yaw rate and sideslip tracking errors e_1 and e_2 as implemented in [6, 43–47].

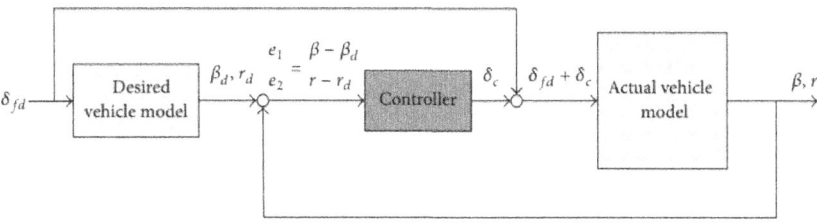

Figure 8: Active front steering control [45].

For control design and analysis of AFS control, the linear state state space model as described in (16) is used. Noted that this equation is similar to equation (12) but the front wheel steer angle, $\delta_f = \delta_{fd} + \delta_c$:

$$\begin{bmatrix} \dot{\beta} \\ \dot{r} \end{bmatrix} = \begin{bmatrix} \dfrac{-C_f - C_r}{mv} & -1 + \dfrac{C_r l_r - C_f l_f}{mv^2} \\ \dfrac{C_r l_r - C_f l_f}{I_z} & \dfrac{-C_f l_f^2 - C_r l_r^2}{I_z v} \end{bmatrix} \begin{bmatrix} \beta \\ r \end{bmatrix}$$

$$+ \begin{bmatrix} \dfrac{C_f}{mv} \\ \dfrac{C_f l_f}{I_z} \end{bmatrix} (\delta_{fd} + \delta_c).$$

(16)

On the other hand, ARS control is used to improve the vehicle response for low speed cornering manoeuvres with the input to the control system being the rear steering angle δ_r. In order to enhance the manoeuvrability at low speed and the handling stability at high speed, combination of AFS control and ARS called 4WAS control has been proposed as implemented in [24, 48, 49]. By implementing 4WAS control, the lateral and yaw motion can be controlled simultaneously using two independent control inputs. Noting that front wheel steer angle δ_f and rear wheel steer angle δ_r with the rear axles of rear tire cornering stiffness, C_r and distance from rear axle to centre of gravity l_r are taken into account in the input matric.

Integrated Active Chassis Control

The integrated active chassis control has become a popular research topic in vehicle dynamics control as discussed in [50]. Vehicle dynamics control can be greatly achieved by integrating the active chassis control of active steering, active braking, and active suspension or active stabiliser as implemented in [12, 23, 51, 52]. Since road-vehicle is usually equipped with front-wheel steering and braking system, an integration and coordination of active front steering and direct yaw moment control are the favourite approaches to achieving the objectives of yaw rate and sideslip control as reported in [2, 10, 11, 27, 28, 53–59]. In this approach, the corrective front wheel steers angle δ_c and corrective yaw moment ΔM_z are considered as two independent control inputs to the vehicle as illustrated in Figure 9.

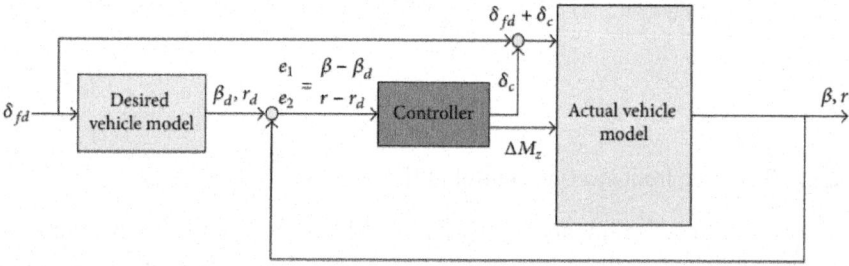

Figure 9: Integrated active front steering-direct yaw moment control [53].

For controller analysis and design of integrated active front steering-direct yaw moment control, the linear state space model used is describe as follows

$$\begin{bmatrix} \dot{\beta} \\ \dot{r} \end{bmatrix} = \begin{bmatrix} \dfrac{-C_f - C_r}{mv} & -1 + \dfrac{C_r l_r - C_f l_f}{mv^2} \\ \dfrac{C_r l_r - C_f l_f}{I_z} & \dfrac{-C_f l_f^2 - C_r l_r^2}{I_z v} \end{bmatrix} \begin{bmatrix} \beta \\ r \end{bmatrix}$$

$$+ \begin{bmatrix} \dfrac{C_f}{mv} & 0 \\ \dfrac{C_f l_f}{I_z} & \dfrac{1}{I_z} \end{bmatrix} \begin{bmatrix} \delta_c \\ \Delta M_z \end{bmatrix} + \begin{bmatrix} \dfrac{C_f}{mv} \\ \dfrac{C_f l_f}{I_z} \end{bmatrix} \delta_{fd}. \tag{17}$$

The principle of active chassis control of steering and braking for yaw stability control has been discussed. From the above discussion, the differences, advantages, and disadvantages of each active chassis control can be digested as tabulated in Table 2. From this table, it can be observed that, by implementing integrated active front steering-direct yaw moment control, the lateral and yaw motions can be controlled simultaneously using two independent control inputs from two different actuators, that is, steering and braking. Thus, this approach could enhance the vehicle yaw stability where the yaw rate and sideslip can be controlled effectively in emergency manoeuvres and the steady state driving condition.

Table 2: Types of active chassis control.

Vehicle actuator	Active chassis control		Advantages	Disadvantages
Brakes	Direct yaw moment control (DYC)	Active braking active differential	(i) Effective for critical driving condition (ii) Good for sideslip/wheelslip control	(i) Less effective for braking on split road surface (ii) Decrease yaw rate during steady state driving condition (iii) Active differential need extra devices
Steering	Active steering control (ASC)	Active front steering (AFS) control	(i) Effective for steady state driving condition (ii) Ease to integrate with braking control (iii) Good for yaw rate control	Less effective during critical driving condition
		Active rear steering (ARS) control	(i) Rear wheel steer angle can be controlled (ii) Good for yaw rate control	Less effective during critical driving condition
		4 wheels active steering (4WAS) control	(i) Two different steer inputs (ii) Good for yaw rate control	Less effective during critical driving condition
Steering and brake	Integrated AFS-DYC control		(i) Two different inputs from two different actuator (steering and braking) (ii) Good for yaw rate and sideslip control	Effective for critical and steady state driving condition

As a conclusion, active chassis control is essential for active yaw stability control system. Therefore, to achieve the yaw stability control objectives, the control strategies for yaw rate and sideslip tracking control are developed based on this active chassis control. The following section will review and discuss the control strategies and algorithms that have been developed in the past.

YAW STABILITY CONTROL STRATEGIES

From the literature, various control strategies have been explored and utilized based on particular algorithm for active yaw stability control such as classical PID controller in [1], LMI based and static state feedback control in [2, 8, 33], H_∞ control theory in [4, 13, 25], sliding mode control (SMC) in [1, 7, 23, 24, 35, 38, 53], optimal guaranteed cost coordination controller (OGCC) in [10], adaptive based control in [11], mixed-sensitivity minimization control techniques in [16], classical controllers PI in [49, 60], internal model control (IMC) in [37], quantitative feedback theory (QFT) in [45], and μ-synthesis control in [48]. Besides that, a combination or integration of different two control schemes to ensure the robustness of yaw stability control has been explored such as SMC and backstepping method in [3], SMC and Fuzzy Logic Control in [12], and LQR with SMC in [17]. As discussed in [20], the IMC and SMC algorithms are designed for yaw stability control and the controllers performances are compared and evaluated. The control strategies are designed based on active chassis control as discussed in Section 4. In active braking or active differential which operates based on direct yaw moment control (DYC), various robust control strategies have been designed. As reported in [3], yaw stability control that consists of tire force observer and cascade controller that is based on sliding mode and backstepping control method is designed. To solve the external disturbance as discussed in [16], the robustness of mixed-sensitivity yaw stability controller is guaranteed for external crosswind and emergency manoeuvres. To cater the uncertainty from longitudinal tire force, the controller for wheel slip control is designed using SMC algorithm for vehicle stability enhancement [17].

As discussed in [20], the second order sliding mode (SOSM) and enhanced internal mode control (IMC) are designed as feedback controller to ensure the robustness against uncertainties and control saturation issues. Both controllers' performances are compared and analysed for yaw control improvement based on rear active differential device. Besides that, the sliding mode control algorithm is also utilized to determine the required yaw moment in order to minimize the yaw rate error and side-slip angle for vehicle stability improvement [22]. To overcome the uncertainties parameters and guarantee robust yaw stability in [25], the control strategy that consists of disturbance observer to estimate feedforward yaw moment and optimal gain-scheduled H_∞ is designed. In the study of [30], the robust yaw moment controller and velocity-dependent state feedback controller are matrixed by solving finite numbers linear matric inequality (LMI).

By using this approach, the designed controller is able to improve the vehicle handling and lateral stability in the presence of uncertainty parameters such as vehicle mass, moment of inertia, cornering stiffness, and variation of road surfaces and also control saturation due to the physical limits of actuator and tire forces. In active steering control, robust control strategies are designed to overcome the uncertainties and external disturbance problems. In [7], adaptive sliding mode control is utilized to estimate the upper bounds of time-derived hyperplane and uncertainties of lateral forces. As discussed in [13], feedback H_∞ control is implemented for robust stabilization of yaw motion where speed and road adhesion variations are considered as uncertainties and disturbance input. As reported in [49], a proportional active front steering control and proportional-integral active rear steering control are designed for four-wheel steering (4WS) vehicle with the objective to overcome the uncertainties of vehicle mass, moment of inertia, and front and rear cornering stiffness coefficients. To ensure a robust stability against system uncertainties, the automatic path-tracking controller of 4WS vehicle based on sliding mode control algorithm is designed [24]. In this study, the cornering stiffness, path radius fluctuation, and crosswind disturbance are considered as uncertainty parameters and external disturbance. As reported in [42], the model reference adaptive nonlinear controllers is proposed for active steering systems to solve the uncertainties and nonlinearities of tire's lateral forces.

Quantitative feedback theory (QFT) technique is implemented for robust active front steering control in order to compensate for the yaw rate response in presence of uncertainties parameters and reject the disturbances [45]. As discussed in [48], robust controller for 4WS vehicle is also designed based on μ-synthesis control algorithm which considers the varying parameters induced by the vehicle during driving conditions as uncertainties while the study in [60] designed the steering control of vision based autonomous vehicle based on the nested PID control to ensure the robustness of the steering controller against the speed variations and uncertainties of vehicle parameters. In integrated active chassis control, an appropriate control scheme is designed to meet the control objectives. Studies in [2, 27, 33] have designed the control scheme that consists of reference model based on linear parameter-varying (LPV) formulation and static-state feedback controller with the objective to ensure the robust performance for integrated active front steering and active differential braking control. In these studies, tire slip angle, longitudinal slips, and vehicle forward speeds are represented as uncertainty parameters.

As reported in [4], integrated robust model matching chassis controller that integrates active rear wheel steering control, longitudinal force compensation, and active yaw moment control is designed using H_∞ controller based on linear matrix inequalities (LMIs) for vehicle handling and lane keeping performance improvement. In integrated active front steering-direct yaw moment control, an optimal guaranteed cost control (OGCC) technique is utilized in [10]. In this study, tire cornering stiffness is treated as uncertainty during variation of driving conditions. As discussed in [11], an adaptive integrated control algorithm based on direct Lyapunov method is designed for integrated active front steering and direct yaw moment control with cornering stiffness is considered as a variation parameter to ensure the robustness of designed controller. As reported in [23], sliding mode controller is utilized for stabilising the forces and moments in integrated control schemes that coordinated the steering, braking, and stabiliser. In this study, the integrated control structure is composed of a main loop controller and servo loop controller that computes and distributes the stabilizing forces/moments, respectively. From the above discussion, these control strategies and algorithms can be summarized and compared in terms of their active chassis control, control objective, advantages, and disadvantages as tabulated in Table 3. In conclusion, an appropriate control strategy must be designed based on particular algorithm. Robust control algorithms such as H_∞, SMC, IMC, OGCC, QFT. are essential to solve the uncertainties and disturbance problems that influenced the yaw stability control performances.

It is revealed that the designed controllers in the above discussion are able to track the desired yaw rate and vehicle sideslip response considering external disturbances and system uncertainty.

Table 3: Yaw stability control algorithms

Control algorithms	Active chassis control	Control objective	Advantages	Disadvantages
PID controller	DYC	sideslip	Anti-wind-up strategy to avoid high overshoot and large settling time	Uncertainties are not consider
LMI static state feedback	Integrated AFS-active differential	Yaw rate and sideslip	robust for uncertainties	
H_∞	Integrated chassis control, active steering	Yaw rate	Robust for uncertainties, reject disturbance	Transient response improvement is not consider
SMC	DYC, active steering	Yaw rate and sideslip	robust for uncertainties and reject disturbance	
OGCC	Integrated AFS-DYC	Yaw rate and sideslip	Robust for uncertainties	
Adaptive integrated control	Integrated AFS-DYC	Yaw rate and sideslip	Robust for uncertainties	
Mixed-sensitivity minimization control	DYC	Yaw rate	Robust for uncertainty, reject disturbance	
PI controller	4WAS	Yaw rate	Robust for uncertainties	
IMC	DYC	Yaw rate	Robust for uncertainty	Transient response improvement is not consider
QFT	AFS	Yaw rate	Robust for uncertainties, reject disturbance	
μ synthesis control	4WAS	Yaw rate and sideslip	Robust for uncertainties	
SMC-backstepping		Yaw rate and sideslip	Robust for nonlinearities	Uncertainties are not considered
SMC-FLC	Integrated steering, brake, and suspension	Yaw rate, sideslip, and roll angle	Robust for uncertainties and nonlinearities	Transient response improvement is not consider
SMC-LQR	DYC	Yaw rate and sideslip	Robust for uncertainty	

YAW STABILITY CONTROL

Problems In the real environments, the dynamics of road-vehicle is highly nonlinear and incorporated with uncertainties. Vehicle motion with nonlinear tire forces represents a nonlinear system where the tire dynamic exhibit nonlinear characteristics, especially during critical driving conditions such as a severe cornering manoeuvre. The main problems of yaw rate and sideslip tracking control are uncertainties caused from variations of dynamics parameters as discussed in the previous section such as road surface adhesion coefficients [8, 13, 33, 37, 45], tire cornering stiffness [2, 8, 10–12, 20, 24, 30, 48, 49], vehicle mass [20, 30, 38, 45, 49], vehicle speed [2, 13, 45], and moment of inertia [30, 49]. Besides that, an external disturbance such as lateral crosswind may influence the tracking control of desired yaw rate and sideslip response as reported in [4, 6, 13, 24].

Therefore, appropriate control strategies and algorithms are essential to overcome these problems as discussed in the previous section. From the view of control system engineering, the transient response performances of tracking control are very important. However, the control strategies and algorithms discussed above are not accommodated for transient response improvement of the yaw rate and sideslip tracking control in presence of uncertainties and disturbances. The designed controllers are only sufficient to track the desired responses in the presence of such problems. Hence, an appropriate control strategy that could improve the transient performance of robust yaw rate and sideslip tracking control should be designed for an active yaw control system which can enhance the vehicle handling and stability performances

HIGH PERFORMANCE ROBUST

Tracking Controller In this section, a principle of possible robust tracking control strategy with high performance that can be implemented for yaw rate and sideslip tracking control is discussed. Based on the literature, a sliding mode control with the nonlinear sliding surface can be proposed to improve the transient response of the yaw rate and sideslip tracking control in presence of uncertainties and disturbances.

Sliding Mode Control (SMC).

Sliding mode control (SMC) algorithm that had been developed in the two last decades is recognized as an effective robust controller to cater for the matched and mismatched uncertainties and disturbances for linear and nonlinear system. It is also utilized as an observer for estimation and identification purpose in engineering system. Various applications using SMC are successfully implemented as numerous research studies and reports have been published. In vehicle and automotive studies, SMC is one of the prominent control algorithms that is used as a robust control strategy as implemented in [3, 17, 38, 53, 61– 63]. Sliding mode control design consists of two important steps, that is, designing a sliding surface and designing the control law so that the system states are enforced to the sliding surface. The design of sliding surface is very important as it will determine the dynamics of the system being control. In conventional SMC, a linear sliding surface has a disadvantage in improving transient response performance of the system due to constant closed loop damping ratio. Therefore, a nonlinear sliding surface that changes a closed loop system damping ratio to achieve high performance of transient response and at the same time ensure the robustness has been implemented in [64–69]. In these studies, the nonlinear sliding surface is designed based on the composite nonlinear feedback (CNF) algorithm.

Nonlinear Sliding Surface Based CNF

The concept of varying closed loop damping ratio, which could improve transient response for uncertain system, is based on composite nonlinear feedback (CNF) control technique. This technique that has been established in [70–74] is developed based on state feedback law. In practice, it is desired that the control system to obtain fast response time with small overshoot. But in fact, most of control scheme makes a tradeoff between these two transient performance parameters. Hence, the CNF control technique keeps low damping ratio during transient and varied to high damping ratio as the output response closed to the set point as illustrated in Figure 10.

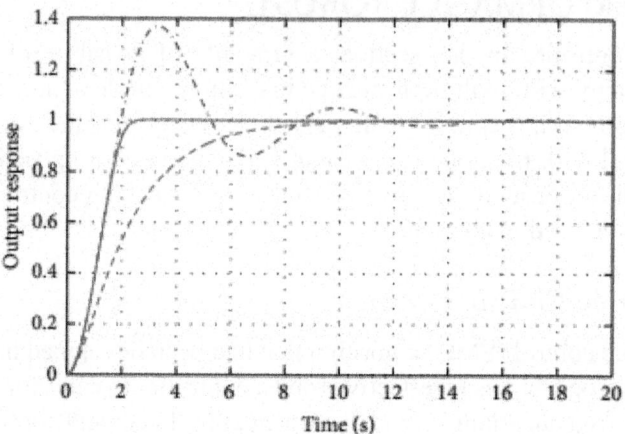

Figure 10: CNF control technique for transient performances improvement [75].

In general, the design of the CNF control technique consists of linear and nonlinear control law as describe as follows:

$$u = [u_{\text{Linear}}] + [u_{\text{Nonlinear}}],$$

$$u = [Fx + Gr] + [\rho(r, y) B'P(x - x_e)], \tag{18}$$

where F is feedback matrix, G is a scalar, B is input matrix, $P>0$ is a solution of Lyapunov equation, and $\rho(r, y)$ is

nonlinear function which is not unique and can be chosen from the following equations:

$$\rho(r, y) = -\beta e^{-\alpha(y-r)^2},$$

$$\rho(r, y) = -\beta e^{-\alpha|y-r|},$$

$$\rho(r, y) = -\frac{\beta}{1 - e^{-1}} \left(e^{-(1-(y-y_0)/(r-y_0))^2} - e^{-1} \right). \tag{13}$$

Based on tracking error, a nonlinear sliding surface adapted from the CNF control law for an active yaw control system can be defined as follows:

$$s := c^T e(t) = [c_1 \quad I_m] \begin{bmatrix} e_1(t) \\ e_2(t) \end{bmatrix}, \tag{20}$$

Where

$$c_1 := F - \rho(r, y) B'P, \tag{21}$$

where $e_1(t)$ and $e_2(t)$ could represent the yaw rate and sideslip tracking error, respectively, B is an input matrix of the system, and I_m is the identity matrix. Then, the nonlinear sliding surface stability can be determined using Lyapunov stability analysis and implement in the designed control law of SMC. Based on the above discussion, the SMC with nonlinear sliding surface based on CNF technique could achieve high performance for uncertain systems. It could improve the transient response performance in the presence of uncertainties and external disturbances. In addition, it is found that this control strategy has not yet been examined for vehicle yaw stability control system and should be further investigated. Therefore, this control technique has initiated a motivation to implement it for robust yaw rate and sideslip tracking control in active yaw control systems. It is expected that this approach could improve the vehicle handling and stability performances.

CONTROLLER EVALUATIONS

In order to evaluate the performance of designing controller, simulations of emergency braking and driving manoeuvres with the nonlinear vehicle model are usually carried out according to ISO or SAE standards. The pure computer simulations, cosimulation with other software or hardware in the loop simulations (HILS), are the common approaches to conducting the yaw stability test with or without driver model for open loop or closed loop analysis, respectively.

One of the typical emergency braking manoeuvres for vehicle yaw stability test is split -μ braking as reported in [2, 37, 60]. In this test, the step input of brake torque is applied to the vehicle in forward motion with constant speed on split road surface adhesion coefficient, μ, where one side of the wheels is on low μ and the other sides of the wheels are on high μ or vice versa. This test is performed to test the vehicle straight ahead driving stability. Critical driving manoeuvres are also another efficient way to test the yaw and lateral stability performances. A step steer manoeuvre can be implemented to evaluate the steady state and transient behavioural response of the vehicle as conducted in [16, 53, 55, 63]. Similarly, the constant speed J-turn manoeuvre is also conducted for such purpose as reported in [5, 8, 9, 15, 30, 33, 45]. Another type of critical driving manoeuvre is lane change manoeuvre as implemented in [3, 5, 10, 11, 15, 20, 21, 23, 26, 45, 46, 53, 55]. This manoeuvre can be conducted for open loop single lane change or closed loop double lanes change with driver model, lane change on different road conditions, lane change on split-μ road, and lane change with braking effect. With steering angle input is in sinusoidal form, the transient handling behaviour can be evaluated and vehicle yaw and lateral stability can be analysed. Another test manoeuvres that can be implemented for yaw stability control are steer reversal test for transient performance evaluation [16, 19, 20], constant speed steering pad to evaluate the steady state vehicle performance [19, 20], steering wheel frequency sweep for the bandwidth, and resonance peak analysis [20] and also fishhook manoeuvre as mentioned in [2, 25, 27]. In order to evaluate the yaw stability control system performance in the presence of disturbance, a crosswind disturbance as reported in [4, 6, 20, 24] is considered as external disturbance that can influence the lateral dynamic stability. During critical driving manoeuvres, the actual response of vehicle's yaw rate and sideslip is obtained and analysed in presence of uncertainties and external disturbances. By performing the test manoeuvres as discussed above, it can be concluded that the ability of the designed controller to track the desired response should be validated. The responses are usually compared to uncontrolled vehicle's responses and other controllers for their steady state and transient response performances.

CONCLUSION

This paper has extensively reviewed the elements of yaw stability control system. In designing yaw stability controller, all these elements, that is, vehicle models, control objectives, active chassis control, and control strategies, play an important role that contributes to the control system performances. For controller design and evaluation, a 2 DOF linear and 7 DOF nonlinear vehicle models are essential. In order to improve the handling and stability performances, the yaw rate and sideslip tracking control are the main objectives that must be achieved by the design controller. To realize an active yaw stability control, an active chassis control of steering, braking or integration of both chassis could be implemented with an appropriate control strategies and algorithms.

In real driving condition, the uncertainties and external disturbance may influenced the yaw rate and sideslip tracking control performances. Hence, the robust control algorithm is necessary. Based on this review, it has been concluded that sliding mode control (SMC) is the best robust controller to address these problems. From the view of control system, transient performances are very important for tracking control. However, an existing SMC configuration does not have capability to improve this transient performance. To address this issue, a nonlinear sliding surface of SMC is designed based on composite nonlinear feedback (CNF) algorithm. This is because the CNF algorithm has been proven in improving transient performances as discussed above. For future works, this control strategy will be implemented for yaw stability control system and the transient performances of yaw rate and sideslip tracking control will be evaluated and compared with classical SMC and other controllers.

CONFLICT OF INTERESTS

The authors declare that there is no conflict of interests regarding the publication of this paper.

ACKNOWLEDGMENTS

The authors would like to thank to Ministry of Education of Malaysia, UTeM, and UTM for the supports of the studies.

REFERENCES

1. B. Lacroix, Z. Liu, and P. Seers, "A comparison of two control methods for vehicle stability control by direct yaw moment," Applied Mechanics and Materials, vol. 120, pp. 203–217, 2012.

2. S. Ç. Baslamisli, I. E. Kose, and G. Anlaş, "Handling stability improvement through robust active front steering and active differential control," Vehicle System Dynamics, vol. 49, no. 5, pp. 657–683, 2011.

3. H. Zhou and Z. Liu, "Vehicle yaw stability-control system design based on sliding mode and backstepping control approach," IEEE Transactions on Vehicular Technology, vol. 59, no. 7, pp. 3674–3678, 2010.

4. J. Wu, Q. Wang, X. Wei, and H. Tang, "Studies on improving vehicle handling and lane keeping performance of closed-loop driver-vehicle system with integrated chassis control," Mathematics and Computers in Simulation, vol. 80, no. 12, pp. 2297–2308, 2010.

5. G. Tekin and Y. S. Ünlüsoy, "Design and simulation of an integrated active yaw control system for road vehicles," International Journal of Vehicle Design, vol. 52, no. 1–4, pp. 5–19, 2010.

6. H. Ohara and T. Murakami, "A stability control by active angle control of front-wheel in a vehicle system," IEEE Transactions on Industrial Electronics, vol. 55, no. 3, pp. 1277–1285, 2008.

7. Y. Ikeda, "Active steering control of vehicle by sliding mode control— switching function design using SDRE," in Proceedings of the IEEE International Conference on Control Applications (CCA '10), pp. 1660–1665, Yokohama, Japan, September 2010.

8. H. Du, N. Zhang, and F. Naghdy, "Velocity-dependent robust control for improving vehicle lateral dynamics," Transportation Research C: Emerging Technologies, vol. 19, no. 3, pp. 454–468, 2011.

9. B. L. Boada, M. J. L. Boada, and V. Díaz, "Fuzzy-logic applied to yaw moment control for vehicle stability," Vehicle System Dynamics, vol. 43, no. 10, pp. 753–770, 2005.

10. X. Yang, Z. Wang, and W. Peng, "Coordinated control of AFS and DYC for vehicle handling and stability based on optimal guaranteed cost theory," Vehicle System Dynamics, vol. 47, no. 1, pp. 57–79, 2009.

11. N. Ding and S. Taheri, "An adaptive integrated algorithm for active front steering and direct yaw moment control based on direct Lyapunov method," Vehicle System Dynamics, vol. 48, no. 10, pp. 1193–1213, 2010.

12. S.-B. Lu, Y.-N. Li, S.-B. Choi, L. Zheng, and M.-S. Seong, "Integrated control on MR vehicle suspension system associated with braking and steering control," Vehicle System Dynamics, vol. 49, no. 1-2, pp. 361–380, 2011.

13. S. Mammar and D. Koenig, "Vehicle handling improvement by active steering," Vehicle System Dynamics, vol. 38, no. 3, pp. 211–242, 2002.

14. C. Zhao, W. Xiang, and P. Richardson, "Vehicle lateral control and yaw stability control through differential braking," in Proceedings of the International Symposium on Industrial Electronics (ISIE '06), pp. 384–389, July 2006.

15. M. Mirzaei, "A new strategy for minimum usage of external yaw moment in vehicle dynamic control system," Transportation Research C: Emerging Technologies, vol. 18, no. 2, pp. 213–224, 2010.

16. V. Cerone, M. Milanese, and D. Regruto, "Yaw stability control design through a mixed-sensitivity approach," IEEE Transactions on Control Systems Technology, vol. 17, no. 5, pp. 1096–1104, 2009.

17. S. Zheng, H. Tang, Z. Han, and Y. Zhang, "Controller design for vehicle stability enhancement," Control Engineering Practice, vol. 14, no. 12, pp. 1413–1421, 2006.

18. E. Esmailzadeh, A. Goodarzi, and G. R. Vossoughi, "Optimal yaw moment control law for improved vehicle handling," Mechatronics, vol. 13, no. 7, pp. 659–675, 2003.

19. M. Canale and L. Fagiano, "Comparing rear wheel steering and rear active differential approaches to vehicle yaw control," Vehicle System Dynamics, vol. 48, no. 5, pp. 529–546, 2010. · ·

20. M. Canale, L. Fagiano, A. Ferrara, and C. Vecchio, "Comparing internal model control and sliding-mode approaches for vehicle yaw control," IEEE Transactions on Intelligent Transportation Systems, vol. 10, no. 1, pp. 31–41, 2009.

21. S. Moon, W. Cho, and K. Yi, "Intelligent vehicle safety control strategy in various driving situations,"Vehicle System Dynamics, vol. 48, no. 1,

pp. 537–554, 2010.

22. S. Yim, W. Cho, J. Yoon, and K. Yi, "Optimum distribution of yaw moment for unified chassis control with limitations on the active front steering angle," International Journal of Automotive Technology, vol. 11, no. 5, pp. 665–672, 2010.

23. D. Li, S. Du, and F. Yu, "Integrated vehicle chassis control based on direct yaw moment, active steering and active stabiliser," Vehicle System Dynamics, vol. 46, no. 1, pp. 341–351, 2008. ··

24. T. Hiraoka, O. Nishihara, and H. Kumamoto, "Automatic path-tracking controller of a four-wheel steering vehicle," Vehicle System Dynamics, vol. 47, no. 10, pp. 1205–1227, 2009. ··

25. S.-H. Yon, O.-S. Jo, S. Yoo, J.-O. Hahn, and K. I. Lee, "Vehicle lateral stability management using gain-scheduled robust control," Journal of Mechanical Science and Technology, vol. 20, no. 11, pp. 1898–1913, 2006.

26. S. H. Tamaddoni, S. Taheri, and M. Ahmadian, "Optimal preview game theory approach to vehicle stability controller design," Vehicle System Dynamics, vol. 49, no. 12, pp. 1967–1979, 2011.

27. S. Ç. Baslamisli, I. E. Köse, and G. Anlaş, "Gain-scheduled integrated active steering and differential control for vehicle handling improvement," Vehicle System Dynamics, vol. 47, no. 1, pp. 99–119, 2009.

28. P. Falcone, H. Eric Tseng, F. Borrelli, J. Asgari, and D. Hrovat, "MPC-based yaw and lateral stabilisation via active front steering and braking," Vehicle System Dynamics, vol. 46, no. 1, pp. 611–628, 2008.

29. W. Cho, J. Yoon, J. Kim, J. Hur, and K. Yi, "An investigation into unified chassis control scheme for optimised vehicle stability and manoeuvrability," Vehicle System Dynamics, vol. 46, no. 1, pp. 87–105, 2008.

30. H. Du, N. Zhang, and G. Dong, "Stabilizing vehicle lateral dynamics with considerations of parameter uncertainties and control saturation through robust yaw control," IEEE Transactions on Vehicular Technology, vol. 59, no. 5, pp. 2593–2597, 2010.

31. Q. Li, G. Shi, J. Wei, and Y. Lin, "Yaw stability control using the fuzzy PID controller for active front steering," High Technology Letters, vol. 16, no. 1, pp. 94–98, 2010.

32. W. J. Manning and D. A. Crolla, "A review of yaw rate and sideslip controllers for passenger vehicles,"Transactions of the Institute of

Measurement and Control, vol. 29, no. 2, pp. 117–135, 2007.

33. S. C. Başlamişli, I. E. Köse, and G. Anlaş, "Design of active steering and intelligent braking systems for road vehicle handling improvement: a robust control approach," in Proceedings of the IEEE International Conference on Control Applications (CCA '06), pp. 909–914, Munich 2006.

34. P. Yih and J. C. Gerdes, "Modification of vehicle handling characteristics via steer-by-wire," IEEE Transactions on Control Systems Technology, vol. 13, no. 6, pp. 965–976, 2005.

35. B. Kwak and Y. Park, "Robust vehicle stability controller based on multiple sliding mode control," inProceedings of the SAE World Congress, SAE 2001-01-10602001, 2001.

36. P. Raksincharoensak, T. Mizushima, and M. Nagai, "Direct yaw moment control system based on driver behaviour recognition," Vehicle System Dynamics, vol. 46, no. 1, pp. 911–921, 2008.

37. M. Canale, L. Fagiano, M. Milanese, and P. Borodani, "Robust vehicle yaw control using an active differential and IMC techniques," Control Engineering Practice, vol. 15, no. 8, pp. 923–941, 2007.

38. M. Canale, L. Fagiano, A. Ferrara, and C. Vecchio, "Vehicle yaw control via second-order sliding-mode technique," IEEE Transactions on Industrial Electronics, vol. 55, no. 11, pp. 3908–3916, 2008.

39. P. Falcone, F. Borrelli, J. Asgari, H. E. Tseng, and D. Hrovat, "Predictive active steering control for autonomous vehicle systems," IEEE Transactions on Control Systems Technology, vol. 15, no. 3, pp. 566–580, 2007.

40. P. Falcone, F. Borrelli, H. E. Tseng, J. Asgari, and D. Hrovat, "Linear time-varying model predictive control and its application to active steering systems: stability analysis and experimental validation,"International Journal of Robust and Nonlinear Control, vol. 18, no. 8, pp. 862–875, 2008.

41. F. Borrelli, P. Falcone, T. Keviczky, J. Asgari, and D. Hrovat, "MPC-based approach to active steering for autonomous vehicle systems," International Journal of Vehicle Autonomous Systems, vol. 3, no. 2–4, pp. 265–291, 2005.

42. Y. Kawaguchi, H. Eguchi, T. Fukao, and K. Osuka, "Passivity-based adaptive nonlinear control for active steering," in Proceedings of the 16th IEEE International Conference on Control Applications (CCA '07), pp. 214–219, October 2007.

43. S. Singh, "Design of front wheel active steering for improved vehicle handling and stability," inProceedings of the SAE Automotive Dynamics & Stability Conference, SAE 2000-01-1619, 2000.

44. W. A. H. Oraby, S. M. El-Demerdash, A. M. Selim, A. Faizz, and D. A. Crolla, "Improvement of vehicle lateral dynamics by active front steering control," in Proceedings of the SAE Automotive Dynamics, Stability & Controls Conference and Exhibition, SAE 2004-01-2081, 2004.

45. J.-Y. Zhang, J.-W. Kim, K.-B. Lee, and Y.-B. Kim, "Development of an active front steering (AFS) system with QFT control," International Journal of Automotive Technology, vol. 9, no. 6, pp. 695–702, 2008.

46. B. Zheng and S. Anwar, "Yaw stability control of a steer-by-wire equipped vehicle via active front wheel steering," Mechatronics, vol. 19, no. 6, pp. 799–804, 2009.

47. Q. Li, G. Shi, and J. Wei, "Yaw stability control using the fuzzy PID controller for active front steering,"High Technology Letters, vol. 16, no. 1, pp. 94–98, 2010.

48. G.-D. Yin, N. Chen, J.-X. Wang, and L.-Y. Wu, "A study on μ -synthesis control for four-wheel steering system to enhance vehicle lateral stability," Journal of Dynamic Systems, Measurement and Control, Transactions of the ASME, vol. 133, no. 1, Article ID 011002, 2011.

49. R. Marino, S. Scalzi, and F. Cinili, "Nonlinear PI front and rear steering control in four wheel steering vehicles," Vehicle System Dynamics, vol. 45, no. 12, pp. 1149–1168, 2007.

50. F. Yu, D.-F. Li, and D. A. Crolla, "Integrated vehicle dynamics control-state-of-the art review," inProceedings of the IEEE Vehicle Power and Propulsion Conference (VPPC '08), pp. 835–840, Harbin, China, September 2008.

51. L. Fei and D. Zhaoxiang, "Integrated control of automotive four wheel steering and active suspenion systems based on unifrom model," in Proceedings of the 9th International Conference on Electronic Measurement and Instruments (ICEMI '09), pp. 3551–3556, Beijing, China, August 2009.

52. S. Zhou, L. Guo, and S. Zhang, "Vehicle yaw stability control and its integration with roll stability control," in Proceedings of the Chinese Control and Decision Conference (CCDC '08), pp. 3624–3629, July 2008.

53. Hu and F. He, "Variable structure control for active front steering and direct yaw moment," inProceedings of the 2nd International Conference

on Artificial Intelligence, Management Science and Electronic Commerce (AIMSEC '11), pp. 3587–3590, Zhengzhou, China, August 2011.

54. Hu and B. Lv, "Study on mixed robust control for integrated active front steering and direct yaw moment," in Proceedings of the IEEE International Conference on Mechatronics and Automation (ICMA '10), pp. 29–33, Xi'an, China, August 2010.

55. Z. He and X. Ji, "Nonlinear robust control of integrated vehicle dynamics," Vehicle System Dynamics, vol. 50, no. 2, pp. 247–280, 2012.

56. Ahn, B. Kim, and M. Lee, "Modeling and control of an anti-lock brake and steering system for cooperative control on split-mu surfaces," International Journal of Automotive Technology, vol. 13, no. 4, pp. 571–581, 2012.

57. Poussot-Vassal, O. Sename, L. Dugard, and S. M. Savaresi, "Vehicle dynamic stability improvements through gain-scheduled steering and braking control," Vehicle System Dynamics, vol. 49, no. 10, pp. 1597–1621, 2011.

58. J. Tjoønnäs and T. A. Johansen, "Stabilization of automotive vehicles using active steering and adaptive brake control allocation," IEEE Transactions on Control Systems Technology, vol. 18, no. 3, pp. 545–558, 2010.

59. Rengaraj and D. Crolla, "Integrated chassis control to improve vehicle handling dynamics performance," in Proceedings of the SAE World Congress and Exhibition, SAE 2011-01-0958, April 2011.

60. R. Marino, S. Scalzi, and M. Netto, "Nested PID steering control for lane keeping in autonomous vehicles," Control Engineering Practice, vol. 19, no. 12, pp. 1459–1467, 2011.

61. T. Shim, S. Chang, and S. Lee, "Investigation of sliding-surface design on the performance of sliding mode controller in antilock braking systems," IEEE Transactions on Vehicular Technology, vol. 57, no. 2, pp. 747–759, 2008.

62. Y. M. Sam, J. H. S. Osman, and M. R. A. Ghani, "A class of proportional-integral sliding mode control with application to active suspension system," Systems and Control Letters, vol. 51, no. 3-4, pp. 217–223, 2004.

63. N. Hamzah, Y. M. Sam, H. Selamat, and M. K. Aripin, "GA-based sliding mode controller for yaw stability improvement," in Proceedings of the 9th Asian Control Conference (ASCC '13), Istanbul, Turkey, 2013.

64. Fulwani, B. Bandyopadhyay, and L. Fridman, "Non-linear sliding

surface: towards high performance robust control," IET Control Theory and Applications, vol. 6, no. 2, pp. 235–242, 2012.

65. B. Bandyopadhyay, F. Deepak, I. Postlethwaite, and M. C. Turner, "A nonlinear sliding surface to improve performance of a discrete-time input-delay system," International Journal of Control, vol. 83, no. 9, pp. 1895–1906, 2010.

66. B. Bandyopadhyay and D. Fulwani, "A robust tracking controller for uncertain MIMO plant using non-linear sliding surface," in Proceedings of the IEEE International Conference on Industrial Technology (ICIT '09), Churchill, Australia, February 2009.

67. B. Bandyopadhyay and D. Fulwani, "High-performance tracking controller for discrete plant using nonlinear sliding surface," IEEE Transactions on Industrial Electronics, vol. 56, no. 9, pp. 3628–3637, 2009.

68. S. Mondal and C. Mahanta, "A fast converging robust controller using adaptive second order sliding mode," ISA Transactions, vol. 51, no. 6, pp. 713–721, 2012.

69. S. Mobayen, V. Johari Majd, and M. Sojoodi, "An LMI-based finite-time tracker design using nonlinear sliding surfaces," in Proceedings of the 20th Iranian Conference on Electrical Engineering (ICEE '12), pp. 810–815, Tehran, Iran, May 2012.

70. Y. He, B. M. Chen, and W. Lan, "On improving transient performance in tracking control for a class of nonlinear discrete-time systems with input saturation," IEEE Transactions on Automatic Control, vol. 52, no. 7, pp. 1307–1313, 2007.

71. Cheng, K. Peng, B. M. Chen, and T. H. Lee, "Improving transient performance in tracking general references using composite nonlinear feedback control and its application to high-speed XY-table positioning mechanism," IEEE Transactions on Industrial Electronics, vol. 54, no. 2, pp. 1039–1051, 2007.

72. Y. He, B. M. Chen, and C. Wu, "Composite nonlinear control with state and measurement feedback for general multivariable systems with input saturation," Systems and Control Letters, vol. 54, no. 5, pp. 455–469, 2005.

73. B. M. Chen, T. H. Lee, K. Peng, and V. Venkataramanan, "Composite nonlinear feedback control for linear systems with input saturation: theory and an application," IEEE Transactions on Automatic Control, vol. 48, no. 3, pp. 427–439, 2003.

74. Z. Lin, M. Pachter, and S. Ban, "Toward improvement of tracking performance—nonlinear feedback for linear systems," International Journal of Control, vol. 70, no. 1, pp. 1–11, 1998.

75. Cheng, B. M. Chen, K. Peng, and T. H. Lee, "A MATLAB toolkit for composite nonlinear feedback control—improving transient response in tracking control," Journal of Control Theory and Applications, vol. 8, no. 3, pp. 271–279, 2010.

Chapter 8

DEVELOPING A HUMAN MACHINE INTERFACE (HMI) FOR INDUSTRIAL AUTOMATED SYSTEMS USING SIEMENS SIMATIC WINCC FLEXIBLE ADVANCED SOFTWARE

[1]Erwin Normanyo, [2]Francis Husinu, [3]Ofosu Robert Agyare

[1][3] Department of Electrical and Electronic Engineering, Faculty of Engineering, University of Mines and Technology, Tarkwa, Ghana

[2] Accra Brewery Limited, Accra, Ghana

ABSTRACT

Most industrial plants are automated without any operator-machine interaction. This makes it difficult for the operator to at a glance know the state of the machines and the immediate steps to undertake in order to clear any anomalies occurring in the plant. In this paper, consideration was given to the design of an HMI for an automated boiler plant which can be operated both manually and automatically by the press of start buttons. The design stages included screen interfacing for the HMI, flowchart development, programming the HMI by assigning tags, integration into Step 7 brand of PLC using ethernet, simulation of the program using "PLCSIM" and the programming codes of the automated boiler plant. The system gives a real-time view of the industrial plant, gives reduction in the troubleshooting time for faults and assures of safety of operating personnel. The designed HMI will be useful to manufacturing industries having industrial automated systems.

INTRODUCTION

Most industrial plants are automated without any interactions between the machines and the operator. This makes it difficult for the operator to at a glance know the state of the machines and the immediate actions and steps to undertake in order to clear any anomalies occurring in the plants. Maximum transparency and flexibility is essential for the operator who works in an environment where processes are becoming more complex and requirements for machine and plant functionality are increasing. The combination of emerging information technologies with traditional condition monitoring systems allows for the continuous running status monitoring for essential equipment as well as comprehensive data processing and centralised resource management [1]. Obvious advantages have moved companies from manual to automated forms of processing resulting in improved operational efficiencies, time savings, increased data accuracy, integrity and security and reduced manual entry costs. Essential to all automatic control mechanisms is the feedback principle, which enables a designer to endow a machine with the capacity for self-correction.

Major developments have occurred in the area of automation in recent years. The offered advantages of industrial automation are quality factor, improved productivity, optimisation of manufacturing operations, reduction of waste and labor costs and replacing humans in tasks done in dangerous environments such as fire, space, volcanoes, nuclear facilities, underwater, amongst others [2, 3]. The demerits of industrial automation cannot be overlooked: These are technology limits, unpredictable development costs, high initial costs and increase in the unemployment rate [3, 4].

HMI-based industrial automation systems have been reported in the literature in recent times. In reference to industrial automation systems in the field of Heating, Ventilation and Air Conditioning (HVAC), Attar et al. [5] designed sensor-enabled cubicles for occupant-centric capture of buildings needed to better serve occupants with improved comfort levels. A prototype system that includes data sensing, data storage and data representation to enhance the state of building performance visualisation using HMI was implemented. The hardware implementation was described where each cubicle was turned into a data-sampling cell distributed throughout a typical office floor. In [6] Building Automation Systems (BAS) where sensors are embedded into HVAC equipment to collect large amounts of data about the overall performance of a building were addressed. Differences however are not addressed at the user level due to coarse sampling and insufficient coverage of individual occupancy zones. Chou et al. [7] and Holmes [8] described a real-time data visualisation scheme on an HMI to create a "personalised and contextaware workplace" within an office space. Sensors were embedded into an office cubicle as a means to communicate various comfort-related factors to the user in real-time. Psychological impact of real-time data display as a potent tool for increasing a user's behavioral awareness was acknowledged. In [9] PLC code automatically generated not only works in ideal conditions when the process or production plant is up and running but also able to handle parts such as interlock logic, safety instructions, start-up and shutdown sequences. Four requirements were established to be able to auto generate PLC code. Volvo Car Corporation [10] created the HMI screens for their cars using Siemens WinCC. Since WinCC and Step 7 are highly integrated

the screens were created in the same project as the PLC code. In [11] automatic generation of PLC programs using automation designer software from Siemens virtual commissioning tools was developed. The purpose was to generate PLC code and HMI screens in an earlier phase of the commissioning and also to be able to reuse information created in Process Simulation software stored in a common database. PC-based automation for plant visualisation using WinCC Flexible and Visual Basic.NET was investigated into by Siemens AG. [12]. Data exchange was between controller WinAC RTX 2010 and visualisation occurs via a shared main memory area, which is provided by the Shared Memory Extension (SMX) interface of the WinAC open development kit. A Simatic HMI IPC477C was used for visualisation with WinCC Flexible. The data exchange between WinAC RTX and WinCC Flexible occurs via Softbus. A graphic surface represents data from the controller using control elements, such as input fields, buttons, graphic objects and animations. Flaspöler et al. [13] surveyed the literature on adequate ergonomic design, including HMI, as a priority for the European Union. This was aimed to raise awareness of the importance of adequate HMI as a vital factor for ensuring workers' occupational safety and health. Complexity of HMI leading to safety and health risks such as increased mental and emotional strain for users were explored. Patiño-Forero et al. [14] designed an elevator group control system using PLC and Destination Control System (DCS). Passenger flow was characterised to be highest in a heavy incoming traffic situation such as the traffic observed in the early hours of the morning. Novelty architecture for an elevator system was proposed by implementing a control strategy based on fuzzy logic using DCS. The architecture of the elevator group control system was designed in an industrial controller using DeviceNet industrial network based instrumentation and the DCS are modeled through HMI located at each floor

Scott, Lance and Burton [15] designed a motion library for induction motors using an induction motor fitted with a 1024 pulse A/B phase incremental encoder, Sinamics G110 standard drive and an S7-200 CPU 224XP PLC with the programming software micro/WIN V4.0 SP6. The design allows the commissioning engineer to graphically view the speed, torque or position data of the move using the status trend function of Micro/WIN. Data was displayed on a touch panel or PC running WinCC Flexible. A simulation model importing existing geometry from Tetra Pak´s CAD-software while logics and kinematics were added using Delmia Automation (DA) software was created by Joakin and Tobias [16].

The model was controlled by a virtual PLC through a virtual HMI. The purpose of the simulation was to discover such geometrical problems as collisions early in the development process. Bozzon et al. [17] proposed an innovative use of a mix of networking standards and software implementation technologies for the design of industrial HMI. The technologies that can be fruitfully used in the implementation of HMI architectures were analysed. The design of a real industrial HMI system that exploits internet communication protocols and Webbased architectures was illustrated. Several advanced features such as application adaptivity, interface personalisation, control remotisation and multi-channel notification were achieved. The resulting platform in terms of performance, reliability and usability was finally evaluated. Free Scale Semiconductor [18] implemented an HMI using the Touch Sensing Software (TSS) library. There were two "Extra Credit" sections. The first shows how to add a second sensor and LED. The second shows how to view, touch, and release events in the CodeWarrior Debugger. The application note and hardware uses the Freescale Touch Sensing Software Evaluation Board (TSSEVB) or Freescale 8-bit MCU. The touch pad uses a touch sensor indicated by a LED.

This paper is aimed at developing an HMI for industrial automated systems which presents processed data to a human operator on a screen and through this the human operator monitors and controls the process. The developed HMI should be able to:

- Communicate with Siemens Simatic Step 7 brand of PLC;
- Give a very realistic view of industrial plant;
- Reduce hardware by replacing many push
- buttons, selector switches and lights;
- Replace humans in tasks done in dangerous
- environments; and
- Allow the operator to start and stop cycles.

MATERIALS AND METHODS

Basic Theory on Human Machine Interface

A Human Machine Interface (HMI) is a device for providing the means of controlling, monitoring, managing and visualizing device processes. With controls and readouts graphically displayed on the screen, the operator can use either external buttons or the touch screen to control the machinery [19]. HMI can also be defined as the interaction between a PLC system and an operator. The interaction is presented by a screen with dynamic icons, figures and text. An operator can monitor the production and control it to a certain level by the help of an HMI panel or PC [20]. As technology systems grow more complex, issues of end-product equipment safety, ease of operation and reducing the risk of human error are becoming extremely important. Designers today know that the operational performance, efficiency and safety of a wide range of systems are closely related to the interaction between humans and machines in other words the HMI. The selection and seamless integration of HMI components such as switch controls, actuators and indicators, are critical to the success of equipment designed for human operation [21]. The goal of interaction between a human and a machine at the user interface is effective operation and control of the machine and feedback from the machine which aids the operator in making operational decisions. The user interface includes hardware and software components. User interfaces exist for various systems and provide a means of input allowing the users to manipulate a system, and output allowing the system to indicate the effects of the users' manipulation.

Poor design of HMI however can lead to bad temper and even to negative health effects such as frustration caused by depression as reported by [22]. The most important features of having HMI are high quality graphics for realistic representations of machinery and processes, alarms, trends, simulation, messaging, animation of equipment based on operator standards, hardware cost reduction and communication such as Serial Port (SP), Data Highway Plus (DHP), ethernet and Dynamic Data Exchange (DDE) [23, 24].

Simatic WinCC Flexible Advanced software is ideal for use as the HMI software in all applications in which operator control and monitoring is required on site whether in production or process automation. WinCC Flexible Advance is Windows-based engineering software created by Siemens for producing HMI screens. The screens are available for use on Simatic HMI operator control and monitoring devices as well as standard PCs [25]. The programming interface of the WinCC Flexible Advanced software basically consists of the project view, property view, work area, tool pallet, output view and object view (Fig. 1).

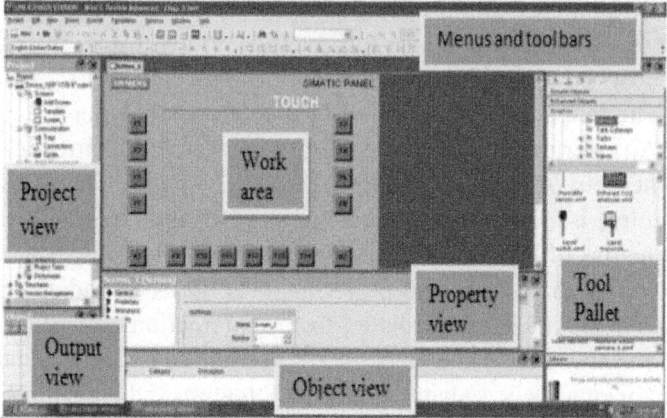

Figure. 2: Gives the block diagram of the design concept of hardware and software.

In terms of design criteria, the HMI should have the following features

- Communicate with Siemens Simatic Step 7 brand of PLC;
- Give a very realistic view of industrial plant;
- Reduce hardware by replacing many push
- buttons, selector switches and lights;
- Replace humans in tasks done in dangerous
- environments; and
- Allow the operator to start and stop cycles.

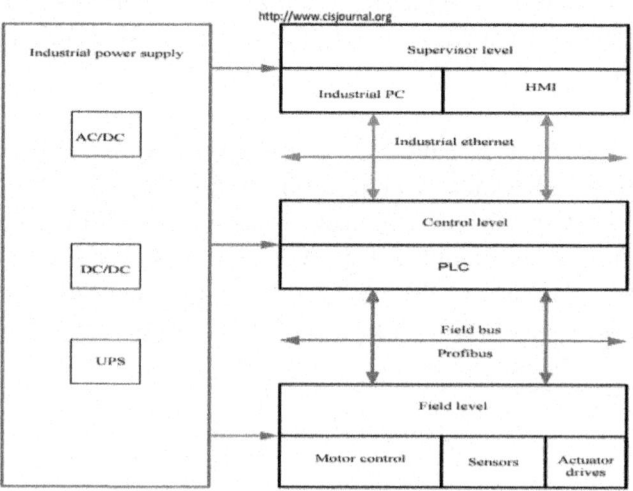

Figure 2: Block Diagram of the Design Concept

In the development of the HMI consideration was given to the design of a typical HMI for a fully automated steam generation boiler plant. The system basically consists of water and diesel reservoirs to feed water and diesel to the boiler. Pumps, digital valves with feedback control and level sensors are used to control the system. When the level of water or diesel inside the reservoir falls to a low level, the level sensors detect and send signals to open the valves so as to pump water or diesel into the reservoir. Also, the condensate water from the steam consumer is fed to a condensate tank and treated to remove hardness from the water. The condensate water after treatment is fed back to the boiler water reservoir to reduce the cost of municipal water supply for steam generation.

Materials

Both hardware and software components are required for the implementation of the HMI for the automated boiler system.

System Hardware

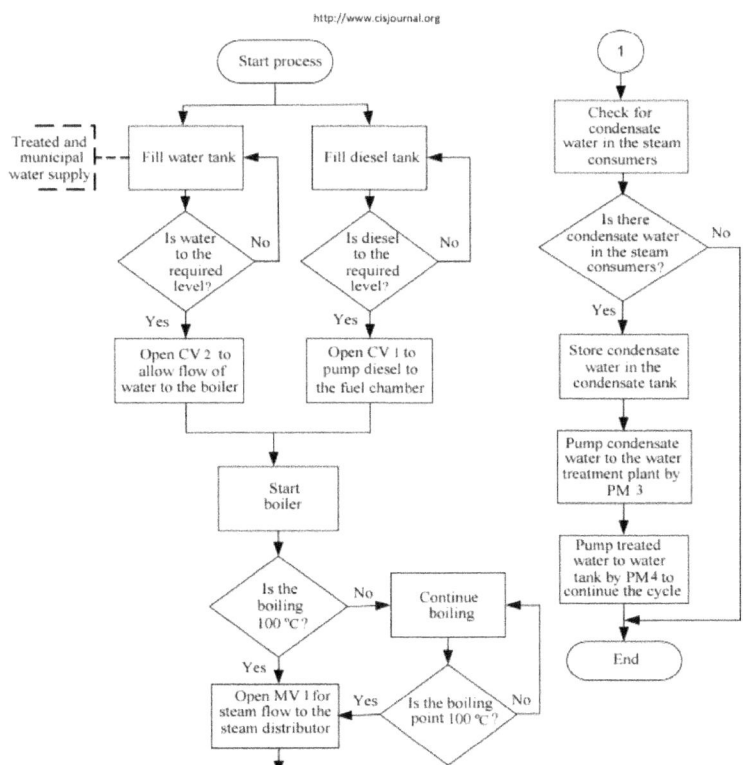

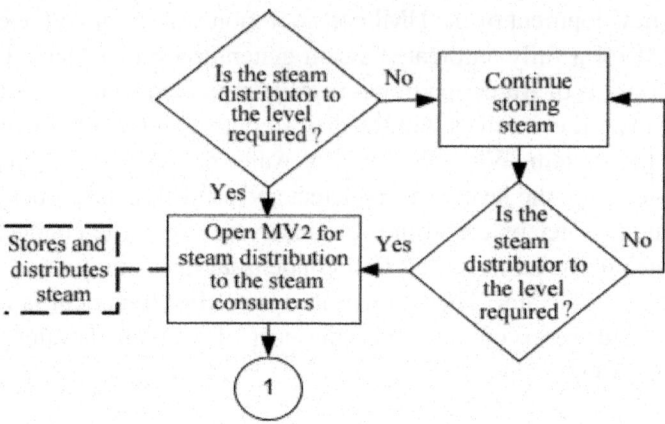

Figure 3: Flowchart for the Operation of the Automated Boiler Plant

The required hardware are desk PC as the HMI screen, industrial ethernet as the link between the HMI and the PLC, profibus to connect the PLC and the field device, CPU 315-2 DP hardware and CP 341 PLC module where automation programs are executed, 24 V DC power supply and field devices such as analog level sensors, PT100, digital valves, non-return valves, water and diesel pumps, pressure transmitters/transducers and pressure relief valves.

System Software Siemens Simatic WinCC

Flexible Advanced and Step 7 software are the main softwares used for the programming of the HMI and the PLC respectively.

Methods

For the methods, flowchart for the operation of the automated boiler plant is developed and presented in Fig. 3. Programming of the HMI is looked at and tags creation is affected. The HMI is then integrated into the industrial ethernet. Finally, simulation of the HMI and the PLC program are conducted using PLCSIM.

Flowchart for the Operation of the Automated Boiler Plant

CV 1 and CV 2 are control valves, MV 1 and MV 2 are motor valves, PM 1, PM 2, PM 3 and PM 4 are pump motors. Any condensate water is stored in the condensate tank. The condensate water is pumped into the water treatment plant by PM 3 to remove any hardness and impurities present to enable the boiler to start boiling at 100 °C. The treated water is pumped into the water tank by PM 4 and the cycle continues as long as the system is in Auto mode. The automated boiler plant is designed to operate in both manual and automatic mode. In manual mode, the operator starts each of the machines sequentially by pressing the switches. Flowchart for the automatic mode is as shown in Fig. 3. The plant is operated by pressing the 'Auto' button after which the "Start Process" button is pressed for the plant to operate in the automatic mode. Operating in the automatic mode means that the protective and sensing elements check the system to ensure that everything is in the required order before initiating the process

Programming the HMI of the Automated Boiler Plant

After generating the flowchart, programming the needed HMI starts with HMI creation and then, design of the automated boiler interface as preliminary stages.

Creating HMI for the automated boiler

To start WinCC Flexible software, either click the desktop icon on the programming device or select it from the windows start menu. It can also be started by: Clicking on start; All programs; Simatic; WinCC Flexible 2008; and WinCC Flexible. With the WinCC Flexible being opened, click on project menu and then click on "new" to create a new project. With this steps completed it will then take you to the device selection window where you select PC option and click the WinCC Flexible Runtime and then click OK in order to create PC based HMI for the boiler plant as shown in Fig. 4.

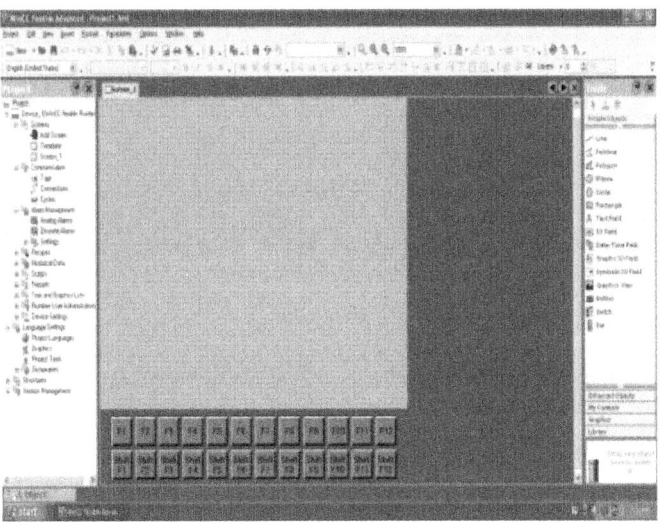

Figure 4: PC Based HMI Window.

Design of the automated boiler interface

In designing the automated boiler interface, the design tool pallet in the right pane is used. The tools pallet contains several tools that can be picked and placed on the working area. They are then joined with the appropriate elements to form the design view. That is, say a pipe will be selected from the tools pallet and used to join the inlet and outlet of say a pipe or a pump to form a physical look of a connected valve or a pump. Fig. 5 depicts the right pane showing the tools pallet. After the design architecture, most of the control elements that initiate the operation, control and monitoring of the automated boiler plant are assigned appropriate tags. Fig. 6 shows control elements on the main screen.

Figure 5: The Right Pane Showing Tools Pallet.

The main screen consists of the control elements such as pipes, control valves, water treatment plant, water tank, diesel tank, municipal water supply, steam consumer, pump motors, steam boiler, condensate tank, steam distributor, and the manual, automatic, stop process and the reset buttons. There are other screens for alarm, warnings, status messages, parameterisation of the control valve in manual mode.

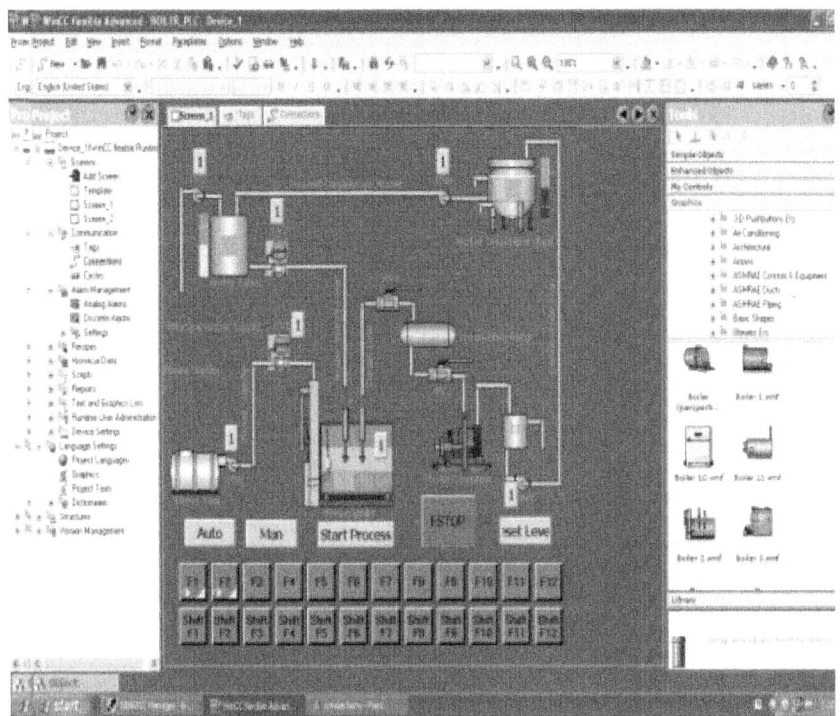

Figure 6: Control Elements on the Main Screen.

Programming the HMI

Programming of the HMI involves assigning tags to screen elements. A tag is basically a link or connection between a screen element of the HMI and an address in the PLC. Tags are created in the tag window. By double clicking tags in WinCC Flexible, the tag window opens. To create a tag: First open the tag window by double clicking on tags in the left pane of the winCC Flexible program; Double click inside the tag window to create a default tag; Edit the default tag at the columns Name, Data Type and Address as shown in Fig. 7

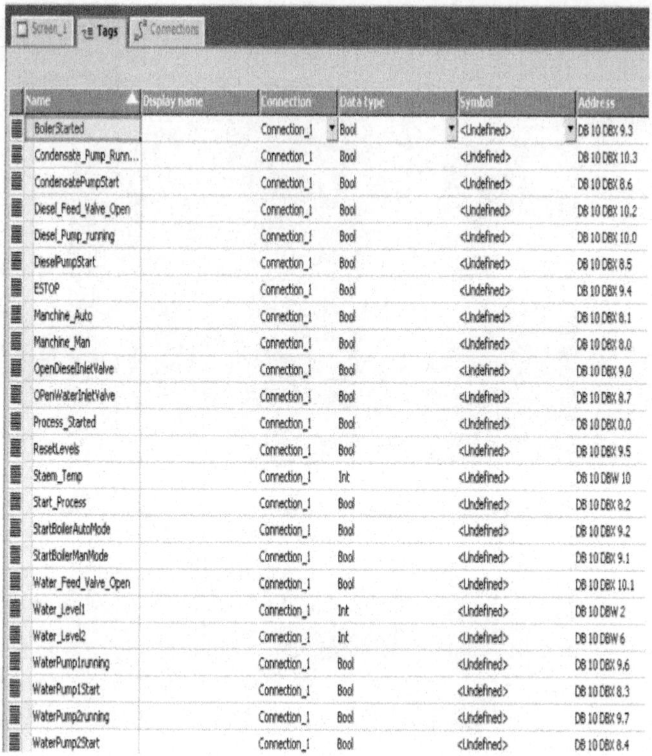

Name	Display name	Connection	Data type	Symbol	Address
BoilerStarted		Connection_1	Bool	<Undefined>	DB 10 DBX 9.3
Condensate_Pump_Runn...		Connection_1	Bool	<Undefined>	DB 10 DBX 10.3
CondensatePumpStart		Connection_1	Bool	<Undefined>	DB 10 DBX 8.6
Diesel_Feed_Valve_Open		Connection_1	Bool	<Undefined>	DB 10 DBX 10.2
Diesel_Pump_running		Connection_1	Bool	<Undefined>	DB 10 DBX 10.0
DieselPumpStart		Connection_1	Bool	<Undefined>	DB 10 DBX 8.5
ESTOP		Connection_1	Bool	<Undefined>	DB 10 DBX 9.4
Manchine_Auto		Connection_1	Bool	<Undefined>	DB 10 DBX 8.1
Manchine_Man		Connection_1	Bool	<Undefined>	DB 10 DBX 8.0
OpenDieselInletValve		Connection_1	Bool	<Undefined>	DB 10 DBX 9.0
OPenWaterInletValve		Connection_1	Bool	<Undefined>	DB 10 DBX 8.7
Process_Started		Connection_1	Bool	<Undefined>	DB 10 DBX 0.0
ResetLevels		Connection_1	Bool	<Undefined>	DB 10 DBX 9.5
Staem_Temp		Connection_1	Int	<Undefined>	DB 10 DBW 10
Start_Process		Connection_1	Bool	<Undefined>	DB 10 DBX 8.2
StartBoilerAutoMode		Connection_1	Bool	<Undefined>	DB 10 DBX 9.2
StartBoilerManMode		Connection_1	Bool	<Undefined>	DB 10 DBX 9.1
Water_Feed_Valve_Open		Connection_1	Bool	<Undefined>	DB 10 DBX 10.1
Water_Level1		Connection_1	Int	<Undefined>	DB 10 DBW 2
Water_Level2		Connection_1	Int	<Undefined>	DB 10 DBW 6
WaterPump1running		Connection_1	Bool	<Undefined>	DB 10 DBX 9.6
WaterPump1Start		Connection_1	Bool	<Undefined>	DB 10 DBX 8.3
WaterPump2running		Connection_1	Bool	<Undefined>	DB 10 DBX 9.7
WaterPump2Start		Connection_1	Bool	<Undefined>	DB 10 DBX 8.4

Figure 7: Created Tags of the Automated Boiler Plant.

Double click the next row of the tag window to create the next tag and so on; At the Address column, assign address in the PLC program to the tag to enable communication between the screen element and the PLC; and Click save to save the tags created.

Tag assignment to screen elements To assign a tag to a screen element (e.g. the screen element labeled "Man");

- Double click on the element say "Man" mode command button. This opens the tag property window;

- Click on the 'Events', option and then select "release" sub-option to the function list window as shown in the right pane. The release option creates a function to be initiated when the button is clicked and then released;

- Double click the first row to open the function list. In the function list, select 'SetBit' to set the tag "On" when the button is clicked and released;

- Double click the row below row 1 in the function list window to open the tag window (opens a window showing all the tags created); and

- Select the tag whose bit is to be set when the "Man" command button is clicked and released. In this case, the "Machine Man" tag (i.e. puts the controls in manual mode) is selected.

- The final window for tag creation is as shown in Fig. 8. The 'ResetBit' function above is assigned to the tag "Machine Auto" to serve as a soft interlock between manual and automatic modes of operation.

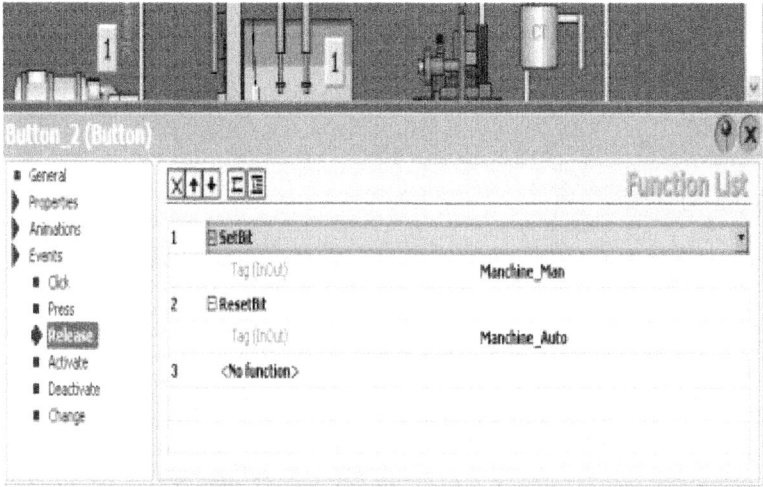

Figure 8: Final Window for Tag Creation.

HMI Integration into Industrial Ethernet

The HMI is integrated into industrial ethernet using Siemens communication processor CP343-1 on the PLC assembly. The communication processor is the highlighted module in Fig. 9. The CP343-1 is assigned an IP address. To assign the CP343-1 a network IP address, the following steps are taken:

- Double click the CP343-1 module to open its property window;

- Click on properties in the above window to open another window;

- Click on new and then click OK to return to the screen in Fig. 9.

- Enter the IP address 192.168.0.1 and Subnet mask of 255.255.255.0;

- Click OK to exit the IP address entering window;

- Click OK again to exit the CP343-1 Properties window;

- Compile the configured hardware; and

- Download the compiled hardware configuration to the PLC through either MPI, ProfiBus DP interface or through industrial Ethernet cable after setting the correct PG interface.

The next stage in the integration process is to create an ethernet connection between the HMI and the Step 7 PLC through CP 343-1. To create an ethernet connection between the HMI and the PLC, open the connection window by double clicking connections in the project window (left pane) in WinCC Flexible. By double clicking the first row in the connections window, a default connection is set. Because the HMI program was integrated into Step 7, the default connection selects S7 300/400 under the communication driver column of the connection window.

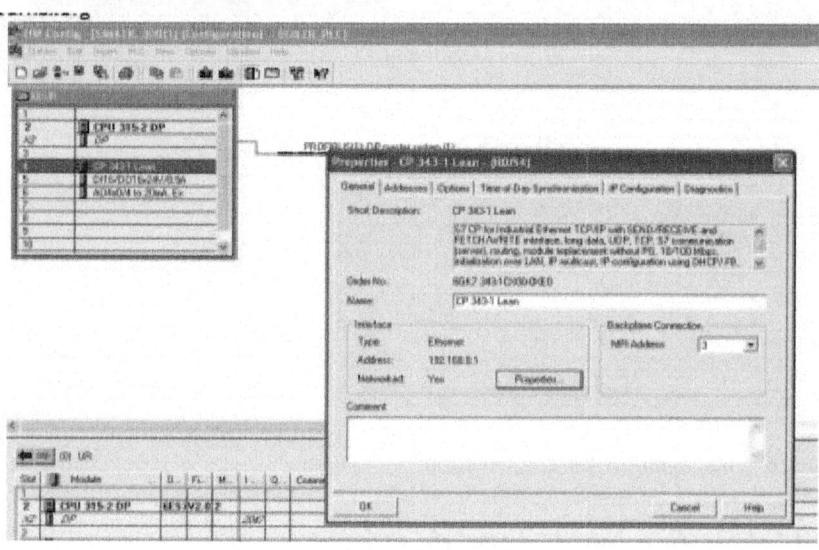

Figure 9: Property Window of CP343-1 Module.

Accept all default entries. The next point is to select the ethernet under interface and then enter the IP address; 192.168.0.1 for the PLC and 192.168.0.2 for the HMI. The steps are as shown in Fig.10.

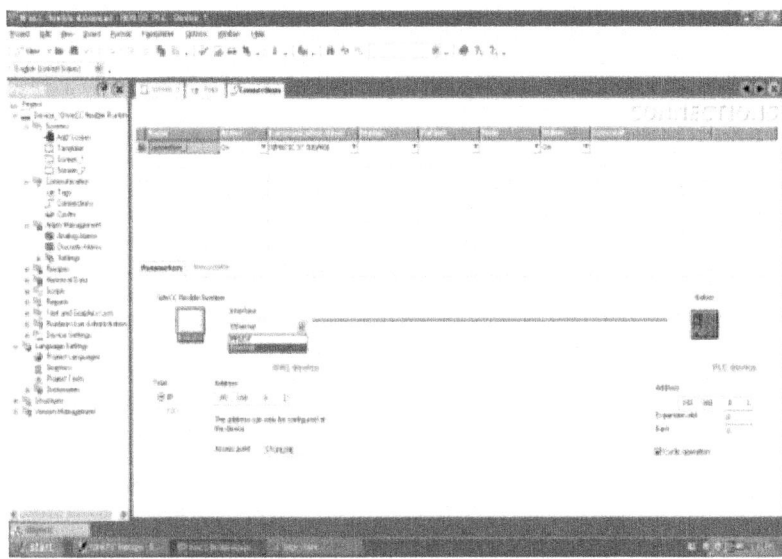

Figure 10: Ethernet Connection between the HMI and the PLC.

Simulation of HMI and PLC Program

Simulation is used to test the response between the HMI program and the PLC program. The Siemens PLC simulator software, PLCSIM is used for the simulation. The simulation process is as follows:

- Open Simatic Manager;
- Click on the PLCSIM icon to open the PLC Simulator;
- Select the appropriate interface (PLCSIM(TCP/IP)) to transfer the PLC program;
- With the S7 PLCSIM opened, go back to Simatic Manager;
- Click Blocks in the left pane to select all the programmed blocks in the right pane. Finally, click on the download icon on the menu bar to download the program to the PLC;
- Switch the PLCSIM CPU into run mode by checking the run option on the CPU window;
- With the PLCSIM in run mode without errors, open the function block FC10 in Simatic Manager and click on 'monitor' to see how the program is running;

- For the next testing process of switching the HMI program also into runtime or simulation mode is achieved by clicking on the runtime or simulation icon shown in Fig. 11; The step 7 launches the HMI program in runtime mode; and

- With the PLC block FC10 in monitoring mode, click the 'Man' command button on the HMI. Check in the PLC programmed block FC10 to see if the tag for the 'Man' command button which is assigned an address DB10.DBX8.0 is activated.

Runtime Icon

Figure 11: Starting WinCC Flexible Runtime

RESULTS AND DISCUSSION

Simulation Results

Testing the functionality of the automated boiler plant is designed in such a way that in runtime when operating in the manual mode, the manual button indicates green and all the other switches on each component indicate red and turn green when the operator presses the switch as shown in Fig. 12. The automatic mode is activated when the 'Auto' button is pressed indicating green. The system starts automatically by performing system checks on the plant before the process is initiated. It then controls itself automatically without any human intervention. Fig. 13 depicts the automatic mode of operation of the automated boiler plant.

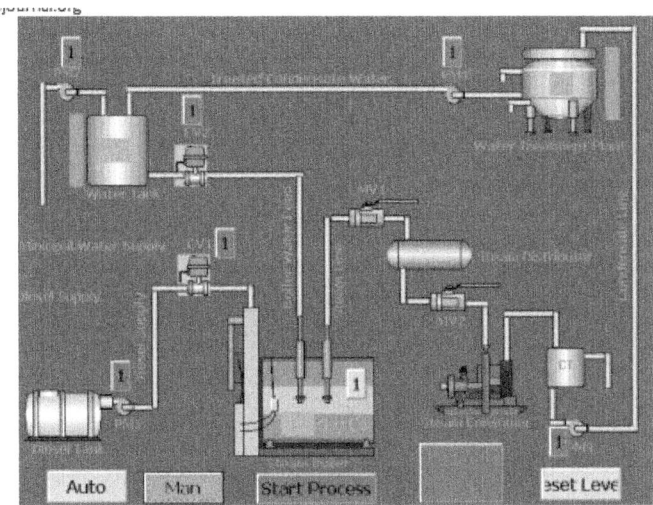

Figure 12: Manual Mode of Operation of the Automated Boiler Plant

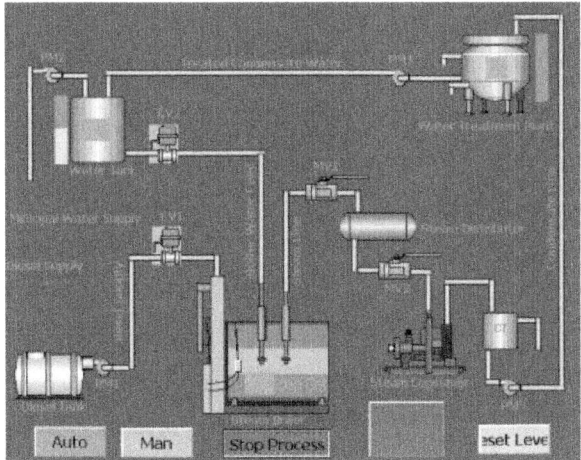

Figure 13: Automatic Mode of Operation of the Automated Boiler Plant

Samples of the related PLC programs are as shown in Fig. 14 and Fig. 15.

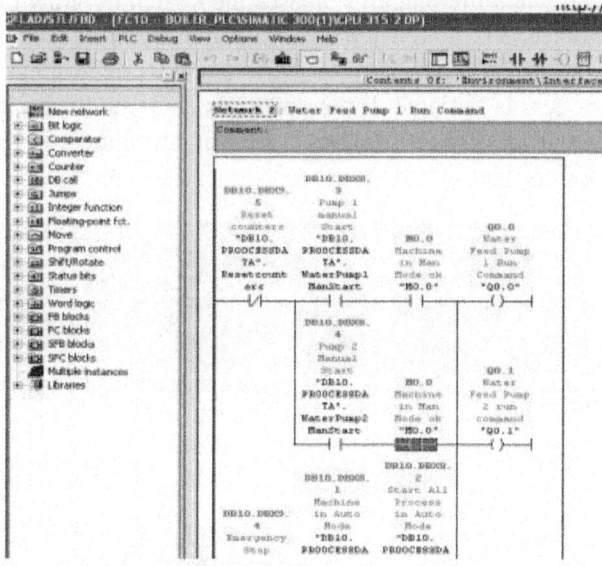

Figure 14: Water Feed Pump 1 Run Command.

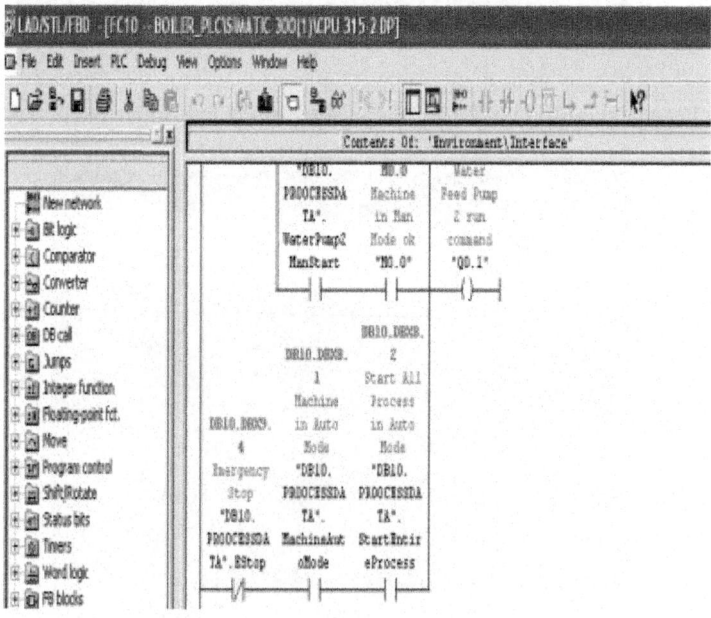

Figure 15: Water Feed Pump 2 Run Command

Discussion of SimulationResults

Successful programming of the designed HMI was as a result of correct tag development for the screen elements to mimic the real-time state of the industrial plant. Successful communication between the HMI and the PLC was trustworthy and facilitated by the tagging system. Manual and automatic modes of operation are availed by way of pressing soft-touch switches. By good design, automatic mode is assured without human intervention.

CONCLUSIONS

The design of HMI using Siemens Simatic WinCC Flexible Advanced software has been achieved. Simulation of the program using "Simatic PLCSIM" was conducted to check the functionality of the PLC program created. Simulation results validated the correctness and applicability of the generated HMI program. The designed HMI is commensurate with: A real-time view of the industrial plant; reduction in the troubleshooting time for faults; and safety of operating personnel. Manufacturing industries can easily adapt to the implementation of HMI as part of their industrial automation systems to make process visualisation of the plant easier, locate faults rapidly and to help replace humans in tasks done in dangerous environments.

REFERENCES

1. L. Wang, K. C. Tan, "Design Principles of Modern Industrial Automation Systems", The Institute of Electrical and Electronics Engineers, Inc, 10pp, 2006.

2. Anon. "Automation", www.altiusdirectory.com/ Business/automation. html, 2010.

3. J. Selender, "Control System Expert VCC", (M. Andersson, & E. Helander, Interviewers) Göteborg, Sweden, 2010.

4. Anon. "Automation", www.en.wikipedia.org/wiki/ Automation, 2009.

5. R. Attar, E. Hailemariam, A. Khan, S. Breslav, G. Kurtenbach, "Sensor Enabled Cubicles for Occupant-Centric Capture of Building Performance Data", MSc Thesis Report, Design Research Associate at Autodesk Research, Toronto, Canada, 8pp, 2005.

6. Z. Mo, A. Mahdavi, "An Agent Based Simulation Assisted Approach to Bilateral Building Systems Control", Proceedings of the IBPSA, Eindhoven, Netherlands, 887pp, 2003.

7. P. M. Chou, J. Gruteser, A. Lai, S., Levas, C. McFaddin, M. Pinhanez, D. Viveros, S. Yoshihama, "Blue Space: Creating a Personalized and

Context-Aware Workspace", IBM Research Report, Yorktown Heights, NY, 98pp, 2001.

8.   T. Holmes, "Eco-visualization: Combining Art and Technology to Reduce Energy Consumption in Creativity and Cognition", Washington, DC, pp. 153-162, 2007.

9.   K. Güttel, P. Weber, A. Fay, "Automatic Generation of PLC Code Beyond the Nominal Sequence", Emerging Technologies and Factory Automation, Hamburg, Germany, pp. 1277 – 1284, 2008.

10.   Volvo Car Corporation," Programming Instructions for PLC Systems Simatic S7", 2008.

11.   M. Andersson, E. Helander, "Automatic Generation of PLC Programs using Automation Designer Based on Simulation Studies and Function Block Libraries", MSc Thesis Report, Production Engineering, Chalmers University of Technology, Göteborg, Sweden, 101pp, 2010.

12.   Siemens AG., "PC-based Automation Plant Visualisation with WinCC Flexible and Visual Basic.NET",www.support.automation.siemens.co m/WW/view/en/, 2011.

13.   E. Flaspöler, A. Hauke, P. Pappachan, R. Dietmar, Bleyer, T. N. Henke, S. Kaluza, A. Schieder, W. Armin, S. Salminen, J.C. Blaise, L. Claudon, J. Ciccotelli, L. Eeckelaert, M. Verjans, K. Muylaert, O.D.B. Rik, "The Human- Machine Interface as an Emerging Risk", The European Agency for Safety and Health at Work, 40pp, 2006.

14.   A. A. Patiño-Forero, D. M. Muñoz, G. Caribé de Carvalho, H. C. Llanos, "Modeling of an Elevator Group Control System using Programmable Logic Control and Destination Control System", ABCM Symposium Series in Mechatronics, University of Brasília, Department of Mechanical Engineering, Brasília, Vol. 4, pp.433- 441, 2010.

15.   W. Scott, B. Lance, B. Burton, "Motion Library for Induction Motor", Abacus Automation, South Africa, 22pp, 2008.

16.   D. Joakim, S. Tobias, "Interactive Control of a Virtual Machine", MSc Thesis Report, Lund University, Department of Industrial Electrical Engineering and Automation, Sweden, 116pp, 2005.

17.   A. Bozzon, M. Brambilla, P. Fraternali, P. Speroni, G. Toffetti, "Applying Web based Networking Protocols and Software Architectures for Providing Adaptivity, Personalisation and Remotisation Features to Industrial Human Machine Interface Applications", MSc Thesis Report, Polytechnic of Milan, Department of Electrical and Information, Italy, 15pp, 2003.

18. Free Scale Semiconductor, "How to Implement a Human Machine Interface using the Touch Sensing Software Library", www. freescalesemiconductor, 2009.

19. Anon. "Industrial Automation Solutions- Human Machine Interface", www.ti.com/general/docs/ge tliteature.tsp, 2011.

20. Siemens AG Automation and Drives, "Operating Manual for WinnCC Flexible 2008 Compact/Standard/Advanced", www.automation .siemens. com/salesmaterialas/brochuresimaticwinc cflble n.pdf,2008.

21. Anon., "Designing Effective Control Interface Systems", www.ewark. com/pdfs/techarticles /ea, 2012.

22. F. Sarodnick, H. Brau,"Methoden der Usability Evaluation", Wissenschaftliche Grundlagen und praktische Anwendung. Huber: Bern, 2006.

23. Anon. (2011b) ,"Krono Tech Instrumentation and Control",www. kronotech.com/HMI/advantages.ht m, 2011.

24. F. Nachreiner, P. Nickel, I. Meyer," Human Factors in Process Control Systems: The Design of Human Machine Interfaces", Safety Science, Vol. 44, pp 5- 26, 2006.

25. Siemens AG Industry Sector, "Simatic WinCC Flexible Brochure March 2010", www.automation. siemens.com/salesmaterial-as/ brochflbleuresimat icwinccn.pdf, 2010.

Chapter 9

MEDICAL ROBOTS: CURRENT SYSTEMS AND RESEARCH DIRECTIONS

Ryan A. Beasley

Department of Engineering Technology and Industrial Distribution, Texas A&M University, TAMU, College Station, USA

ABSTRACT

First used medically in 1985, robots now make an impact in laparoscopy, neurosurgery, orthopedic surgery, emergency response, and various other medical disciplines. This paper provides a review of medical robot history and surveys the capabilities of current medical robot systems, primarily focusing on commercially available systems while covering a few prominent research projects. By examining robotic systems across time and disciplines, trends are discernible that imply future capabilities of medical robots, for example, increased usage of intraoperative images, improved robot arm design, and haptic feedback to guide the surgeon.

INTRODUCTION

Medical robotics is causing a paradigm shift in therapy. The most widespread surgical robot, Intuitive Surgical's da Vinci system, has been discussed in over 4,000 peer-reviewed publications, was cleared by the United States' Food and Drug Administration (FDA) for multiple categories of operations, and was used in 80% of radical prostatectomies performed in the U.S. for 2008, just nine years after the system went on the market [1–3]. The rapid growth in medical robotics is driven by a combination of technological improvements (motors, materials, and control theory), advances in medical imaging (higher resolutions, magnetic resonance imaging, and 3D ultrasound), and an increase in surgeon/patient acceptance of both laparoscopic procedures and robotic assistance. New uses for medical robots are created regularly, as in the initial stages of any technology-driven revolution.

In 1979, the Robot Institute of America, an industrial trade group, defined a robot as "a reprogrammable, multifunctional manipulator designed to move materials, parts, tools, or other specialized devices through various programmed motions for the performance of a variety of tasks." Such a definition leaves out tools with a single task (e.g., stapler), anything that cannot move (e.g., image analysis algorithms), and nonprogrammable mechanisms (e.g., purely manual laparoscopic tools). As a result, robots are generally indicated for tasks requiring programmable motions, particularly where those motions should be quick, strong, precise, accurate, untiring, and/or via complex articulations. The downsides generally include high expense, space needs, and extensive user training requirements. The greatest impact of medical robots has been in surgeries, both radiosurgery and tissue manipulation in the operating room, which are improved by precise and accurate motions of the necessary tools. Through robot assistance, surgical outcomes can be improved, patient trauma can be reduced, and hospital stays can be shortened, though the effects of robot assistance on long-term results are still under investigation.

Medical robots have been reviewed in various papers since the 1990s [4–7]. Many such reviews are domain-specific, for example, focusing on surgical robots, urological robots, spine robots, and so forth [8–13]. For an overview of the basic science behind medical robots (e.g., kinematics, degrees of freedom, ergonomics, and telesurgery) along with a discussion of urologic robotic systems, see Challacombe and Stoianovici [14]. Similarly focused on surgery, Kenngott et al. provide a recent Medline metareview on the outcomes of laparoscopic robot-assisted surgeries (urologic, gynecologic, and abdominal) [15], while Gomes covers market drivers and roadblocks [16], and Okamura et al. explore big picture issues like societal drivers, quantitative diagnosis, and system adaptation/learning [17]. The most recent coverage of medical robots across various domains was by Najarian et al. and the articles collected by Rosen et al. [18, 19].

This paper provides an overview of the impact of robots in multiple medical domains. This work builds on top of the aforementioned papers by providing an updated review of various robotic systems, covering system improvements (technical and regulatory) and changes in manufacturers due to corporate buyouts. Furthermore, to the author's knowledge this work covers more breadth in the medical domains benefiting from robot assistance than any other single paper, and thus provides a big picture view of how robots are improving the medical field. Though primarily focused on commercially available medical robotic systems and the history that describes their evolution, this paper also covers multiple next-generation systems and discusses their potential impacts on the future of the medical field.

NEUROLOGICAL

Brain surgery involves accessing a buried target surrounded by delicate tissue, a task that benefits from the ability for robots to make precise and accurate motions based on medical images [18, 20, 21]. Thus, the first published account investigating the use of a robot in human surgery was in 1985 for brain biopsy using a computed tomography (CT) image and a stereotactic frame [22]. In that work, an industrial robot defined the trajectory for a biopsy by keeping the probe oriented toward the biopsy target even as the surgeon manipulated the approach. This orientation was determined by registering a preoperative CT with the robot via fiducials on a stereotactic frame attached to the patient's skull. That project was discontinued after the robot company was bought out, due to safety concerns of the new owning company, which specified that the robot arm (54 kg and capable of making 0.5 m/s movements) was only designed to operate when separated by a barrier from people. Then in 1991, the Minerva robot (University of Lausanne, Switzerland) was designed to direct tools into the brain under real-time CT guidance. Real-time image guidance allows tracking of targets even as the brain tissue swells, sags, or shifts due to the operation. Minerva was discontinued in 1993 due to the limitation of single-dimensional incursions and its need for real-time CT [23].

The currently available neurosurgery robots exhibit a purpose similar to historical systems, namely, image-guided positioning/orientation of cannulae or other tools (Figure 1). The NeuroMate (by Renishaw, previously by Integrated Surgical Systems, previously by Innovative Medical Machines International) has a Conformité Européenne (CE) mark and is currently used in the process for FDA clearance (the previous generation was granted FDA clearance in 1997) [24]. In addition to biopsy, the system is marketed for deep brain stimulation, stereotactic electroencephalography, transcranial magnetic stimulation, radiosurgery, and neuroendoscopy. Li et al. report in-use accuracy as submillimeter for a frame-based configuration, the same level of application accuracy as bone-screw markers with infrared tracking, and an accuracy of 1.95 mm for the frameless configuration [25].

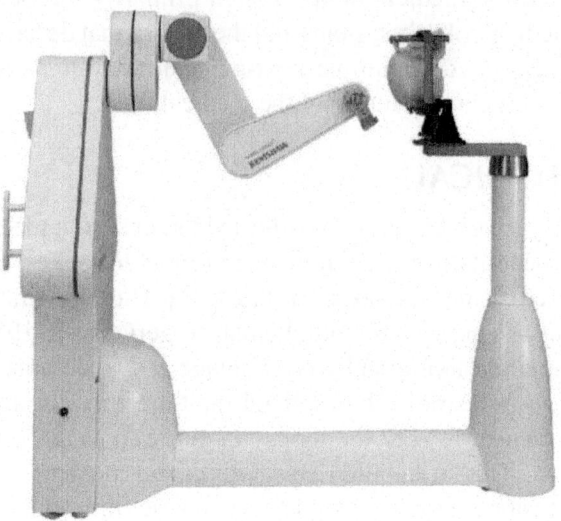

(a) NeuroMate by Renishaw

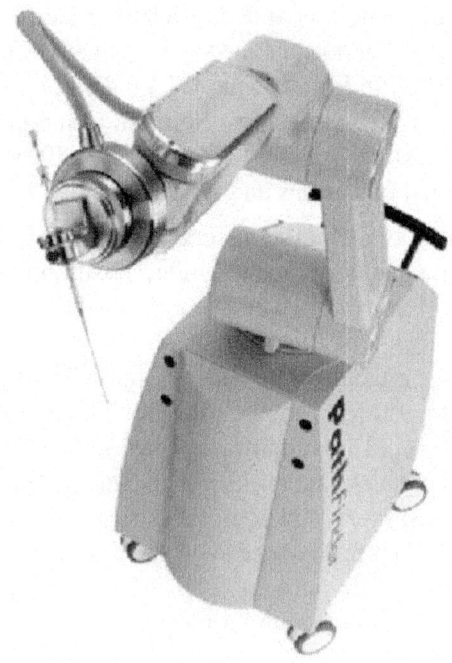

(b) Pathfinder by Prosurgics

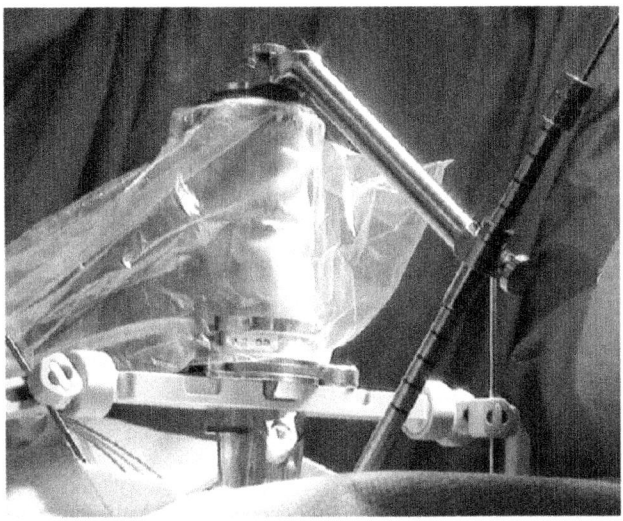

(c) Renaissance by Mazor Robotics

Figure 1: Neurosurgery robots for image-guided tool positioning/orientation. The NeuroMate image is ©2012 Renishaw. The Pathfinder image is ©2012 Prosurgics. The Renaissance image is ©2011 Mazor Robotics Ltd. All rights reserved with non-exclusive permission.

Another robotic system, Pathfinder (Prosurgics, formerly Armstrong Healthcare Ltd.), has been cleared by the FDA for neurosurgery (2004) [26]. Using the system, the surgeon specifies a target and trajectory on a pre-operative medical image, and the robot guides the instrument into position with submillimeter accuracy [27]. Reported uses of the system include guiding needles for biopsy and guiding drills to make burr holes [28].

Renaissance (Mazor Robotics, the first generation system was named SpineAssist) has FDA clearance (2011) and CE mark for spinal surgery, and a CE mark for brain operations (2011) [29]. The device consists of a robot the size of a soda can that mounts directly onto the spine and provides tool guidance based on planning software for various procedures including deformity corrections, biopsies, minimally invasive surgeries, and electrode placement procedures. Renaissance includes an add-on for existing fluoroscopy C-arms that provides 3D images for intraoperative verification of implant placement. Studies show increased implant accuracy and provide evidence that the Renaissance/SpineAssist may allow significantly more implants to be placed percutaneously [30].

ORTHOPEDICS

The expected benefit of robot assistance in orthopedics is accurate and precise bone resection [31, 32]. Through good bone resection, robotic systems (Figure 2) can improve alignment of implant with bone and increase the contact area between implant and bone, both of which may improve functional outcomes and implant longevity [5]. Orthopedic robots have so far targeted the hip and knee for replacements or resurfacing (the exception being the Renaissance system in Section 2 and its use on the spine). Initial systems required the bones to be fixed in place, and all systems use bone screws or pins to localize the surgical site.

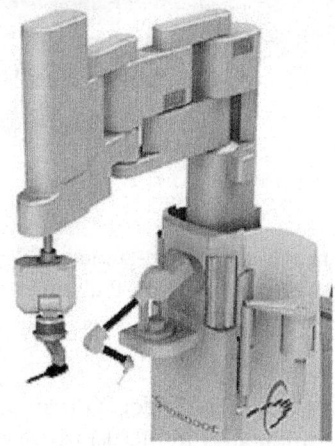

(a) Robodoc by Curexo Technology Corp.

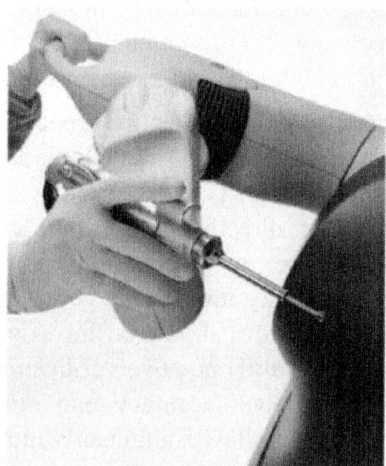

(b) RIO by MAKO Surgical Corp.

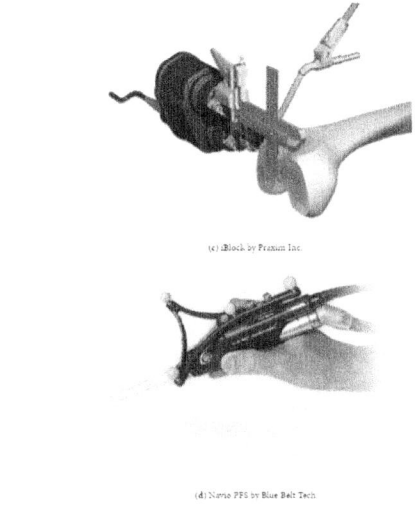

(c) iBlock by Praxim Inc.

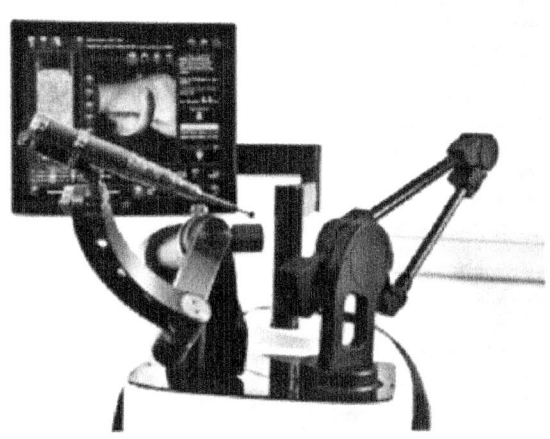

(d) Navio PFS by Blue Belt Tech

(e) Stanmore Sculptor by Stanmore Implants

Figure 2: Orthopedic robots for accurate bone resection. The Robodoc image is ©2012 Curexo Technology Corp. The RIO image is ©2012 MAKO Surgical Corp. The iBlock image is ©2012 Praxim Inc. The Navio PFS image is ©2012 Blue Belt Tech. The Stanmore Sculptor image is ©2012 Stanmore Implants.

The initial robot assistance for orthopedics came via Robodoc (Curexo Technology Corp, originally by Integrated Surgical Systems), first used in 1992 for total hip replacement [5, 33]. Robodoc has received a CE mark (1996), and FDA clearance for total hip replacement (1998) and total knee replacement (2009) [34]. The robot is used in conjunction with OrthoDoc, a surgical planner, with which the surgeon plans bone milling is based on preoperative CT. During the procedure, the patient's leg is clamped to the robot's pedestal, and a second clamp locates the femoral head to automatically halt the robot if the leg moves. The Robodoc then performs the milling automatically based on the surgical plan. Many initial attempts in surgical robotics involved such autonomous motions, which generated concerns about patient and doctor safety. To address those concerns, Robodoc has force sensing on all axes, as well as a six-axis force sensor at the wrist [35]. The force sensing is used for safety monitoring, to allow the surgeon to manually direct the robot arm and to vary the velocity of tool motion as a function of the forces experienced during the milling operation.

Though no longer for sale, CASPAR (Computer Assisted Surgical Planning and Robotics) was another robotic system for knee and hip surgery, introduced in 1997 by OrtoMaquet, acquired by Getinge in 2000, acquired and discontinued by Universal Robot Systems (URS) in 2001. The robot was a direct competitor to Robodoc. It automatically performed bone drilling from a preoperative plan based on CT data.

In 2008, the RIO robotic arm (MAKO Surgical Corp, previous generation called the Tactile Guidance System) was released and received FDA clearance. The RIO is used for implantation of medial and lateral unicondylar knee components, as well as for patellofemoral arthroplasty [36, 37]. As part of the trend away from autonomous robot motions, both the RIO and the surgeon simultaneously hold the surgical tool, with which the surgeon moves about the surgical site. The arm is designed to be low friction and low inertia, so that the surgeon can easily move the tool, backdriving the arm's joint motors in the process [19]. The arm's purpose is to act as a haptic device during the milling procedure, resisting motions outside of the planned cutting envelope by pushing back on the surgeon's hand. Unlike other orthopedic systems, the RIO does not require the bone to be fixed in place, instead relying on a camera system to track bone pins and tools intraoperatively and instantaneously registering the planned cutting envelope to the patient in the operating room. With this configuration, the system has promise for use as a surgical training tool.

Further reducing robotic influence on the cutting tool, the iBlock (Praxim Inc., an Orthopaedic Synergy Inc. company, previous generation the Praxiteles, FDA clearance 2010) is an automated cutting guide for total knee replacement [38]. The iBlock is mounted directly to the bone, preventing any relative motion between the robot and the bone and aligns a cutting guide that the surgeon uses to manually perform planar cuts based on a preoperative plan. Koulalis et al. report reduced surgical time and increased cut accuracy compared with freehand navigation of cutting blocks [39].

The Navio PFS (Blue Belt Technologies, CE mark 2012) does not require a CT scan for unicondylar knee replacement, instead it uses intraoperative planning [40, 41]. The drill tool is tracked during the procedure, and the drill bit is retracted when it would leave the planned cutting volume. Limited information is available on the system due to its recent development.

The Stanmore Sculptor (Stanmore Implants, previous generation the Acrobot Sculptor by Acrobot Company Ltd.) is a synergistic system similar to the RIO, with active constraints to keep the surgeon in the planned workspace [42]. The company's "Savile Row" system tailors a personalized unicondylar knee implant to the patient, incorporates the 3D model of that implant into the surgical planning interface, and uses active constraints with the Stanmore Sculptor to ensure proper preparation of the bone surface. The system does not currently have FDA clearance, but has been in use in Europe since 2004.

GENERAL LAPAROSCOPY

Prior to the 1980s, surgical procedures were performed through sizable incisions through which the surgeon could directly access the surgical site. In the late 1980s, camera technology had improved sufficiently for laparoscopy (a.k.a. minimally invasive surgery), in which one or more small incisions are used to access the surgical site with tools and camera [43]. Laparoscopy significantly reduces patient trauma in comparison with traditional "open" procedures, thereby reducing morbidity and length of hospital stay, but at the cost of increased complexity of the surgical task. Compared with open surgery, in laparoscopy the surgeon's feedback from the surgical site is impaired (reduced visibility and cannot manually palpate the tissue) and tool control is reduced ("mirror-image" motions due to fulcrum effect and loss of degrees of freedom in tool orientation) [16, 44, 45].

Robot assistance for soft-tissue surgery was first done in 1988 using an industrial robot to actively remove soft tissue during transurethral resection of the prostate [5]. As with neurosurgery, the researchers deemed use of an industrial robot in the operating room to be unsafe. The experience provided the impetus for a research system, Probot, with the same purpose [46].

Zeus

Commercial robotic systems for laparoscopy started with Computer Motion's Aesop (discontinued, FDA clearance 1993) for holding endoscopes [47]. Aesop was clamped to the surgical table or to a cart, and either moved the endoscope under voice control or allowed the endoscope to be manually positioned. In 1995, Computer Motion combined two tool-holding robot arms with Aesop to create the Zeus system (discontinued, FDA clearance 2001) [48]. The Zeus's tool arms were teleoperated, following motions the surgeon made with instrument controls (a.k.a. "master" arms or joysticks) at the surgeon console. Technically, the Zeus is not a robot because it does not follow programmable motions, but rather is a remote computer-assisted telemanipulator with interactive robotic arms. To improve precision in tool motion, the Zeus filters out hand tremor, and can scale large hand motions by the surgeon down to short and precise motions by the tool. As described by Marescaux et al., the Zeus was used in the Lindbergh Operation, the first surgery was (cholecystectomy) performed with the surgeon and patient being separated by a distance of several thousand kilometers [49].

Da Vinci

Meanwhile, Intuitive Surgical Inc. was developing the da Vinci (initial FDA clearance 1995, Figure 3(a)). Like the Zeus, the da Vinci is a teleoperated system, wherein the surgeon manipulates instrument controls at a console and the robot arms follow those motions with motion scaling and tremor reduction. Also like the Zeus, the da Vinci was initially offered with three arms to hold two tools and an endoscope, which are mounted to a single bedside cart.

The da Vinci system provides several technical enhancements over the Zeus. The grasper tools have two degrees of freedom inside the patient, the EndoWrist (Figure 3(b)), an enhanced articulation that increases the ease of suturing and other complex manipulations. The console puts increased emphasis on surgeon ergonomics and incorporates a separate video screen for each eye to display 3D video from the 3D endoscope. The motions of the surgeon's hands are mapped to motions of the operational ends of the tools, providing a more intuitive control than the "mirror-image" laparoscopic mapping. In 2003, Intuitive Surgical began selling a fourth arm for the da Vinci, and Intuitive Surgical and Computer Motion were merged (discontinuing the Zeus).

The da Vinci system is the only surgical robot with over a thousand systems installed worldwide and has been sold in four models so far: Standard (1999), S (2006), Si (2009), and Si-e (2010) [50, 51]. The S model increased the image resolution, redesigned the patient-side manipulators to enable multiquadrant access, and shortened setup time. The Si model further improved the visual resolution, refined the instrument controllers, and increased the ergonomics and ease for the surgeon to provide input to the system. The Si-e model is a 3-arm system that is fully upgradeable to the Si model. Continuing the da

Vinci focus on improved visualization, the Firefly Fluorescence Imaging add-on product combines fluorescent dye and a special endoscope to identify vasculature beneath the tissue surface.

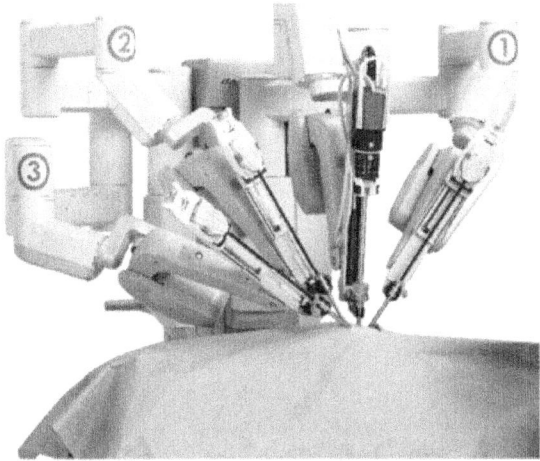

(a) Da Vinci Si patient-side cart

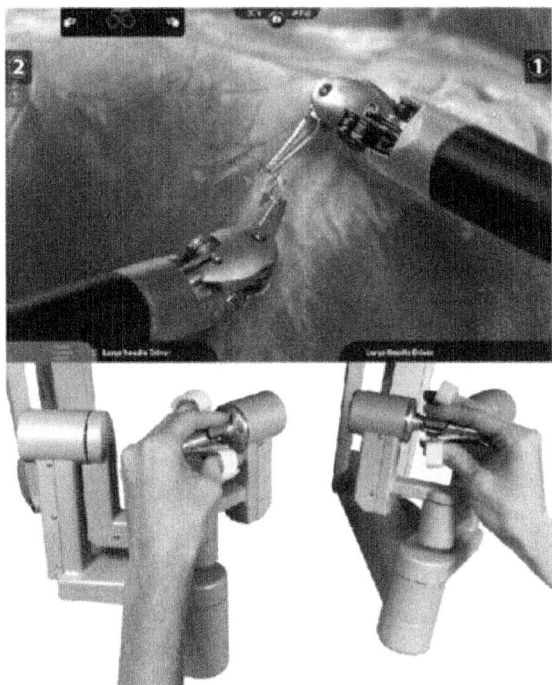

(b) Da Vinci EndoWrist and controllers

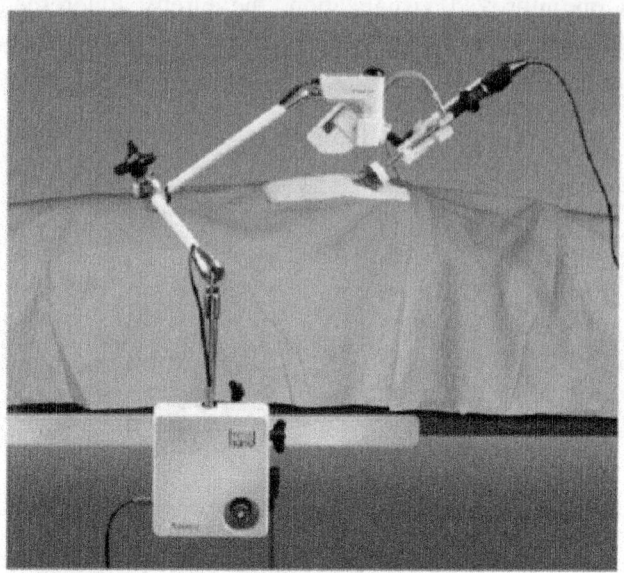

(c) FreeHand by Freehand 2010 Ltd.

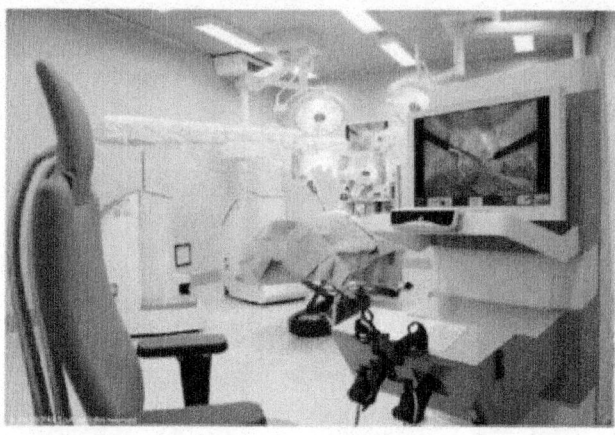

(d) Telelap ALF-X by SOFAR S.p.A

Figure 3: Laparoscopy robots. The da Vinci Si, by Intuitive Surgical Inc. (a) Cart and (b) image mosaic showing the tool tips with EndoWrist articulation, and the instrument controls. (c) FreeHand, a next-generation endoscope holder. (d) A computer model of the Telelap ALF-X, by SOFAR S.p.A. The da Vinci images are ©2012 Intuitive Surgical, Inc. The FreeHand image is ©2012 Freehand 2010 Ltd. The Telelap ALF-X image is ©2012 SOFAR S.p.A. All rights reserved.

The da Vinci was initially cleared for general laparoscopy, became commonly used for radical prostatectomy, and is now cleared by the FDA for various procedures [52, 53]. Even so, as with most or even all robotic systems, long-term benefits continue to be uncertain [15, 54]. The enhanced endoscopic visualization and increased tool articulation are commonly considered improvements, but detractors point out the system's expense (between \$1 M and \$2.3 M), the reduced patient access due to the amount of space the arms take over/around the patient, and the significant amount of training necessary for the best outcomes [55, 56]. To address this last point, the Si model also allows dual console use for training and collaboration, in which both consoles get the same images and can cooperatively control the instruments [57]. Additionally, the da Vinci Skills Simulator is an add-on case that can be used with an Si or Si-e console to practice operations in a virtual environment [58].

In an attempt to further reduce patient trauma, surgeons are exploring Single-Port Access (SPA), LaparoEndoscopic Single-Site surgery (LESS), and Natural Orifice Transluminal Endoscopic Surgery (NOTES) [59, 60]. To meet this need, Intuitive Surgical has recently developed the Single-Site platform for the da Vinci Si model. The Single-Site platform passes two semirigid tools and the endoscope through a single multichannel port, reducing the number of incisions but preventing EndoWrist articulation [61].

FreeHand

The FreeHand robot (Freehand 2010 Ltd., previously Freehand Surgical, previously Prosurgics, the previous generation was called EndoAssist, FDA clearance and CE mark 2009) is a next-generation endoscope holder. The arm (Figure 3(c)) is more compact, easier to setup, and cheaper than its predecessor. Furthermore, endoscope motion is controlled by gentle head motions by the surgeon, which are tracked with an optical system.

Telelap ALF-X

SOFAR S.p.A has developed Telelap ALF-X (CE mark 2011, Figure 3(d)), a four-armed surgical robotic system, to compete with the da Vinci [62]. The system uses eyetracking to control the endoscopic view and to enable activation of the various instruments. Compared to the da Vinci, the system moves the base of the manipulators away from the bed (about 80 cm) and has a realistic tactile-sensing capability due to a patented approach to measure tip/tissue forces from outside the patient, with a sensitivity of 35 grams. The system has been used in animal trials demonstrating a significant reduction in the time for cholecystectomy compared with a "conventional telesurgical system" [62].

PERCUTANEOUS

Noncatheter percutaneous procedures employ needles, cannulae, and probes for biopsy, drainage, drug delivery, and tumor destruction. During the procedure, accurate targeting can be reduced by soft tissue displacements that occur due to patient breathing, changes in posture, or tissue forces exerted during the insertion. Two options to guide a needle to its target are tissue modeling for needle steering and three-dimensional intraoperative imaging [63]. Unfortunately, tissue modeling is excessively complex [64]. So following the latter approach, InnoMotion (Synthes Inc., previously by Innomedic GmbH, CE mark 2005) is a robot arm designed to operate within a CT or magnetic resonance imaging (MRI) machine [65–67]. For MRI-compatibility, the arm (Figure 4(a)) is pneumatically actuated and joint sensing is via MRI-compatible encoders.

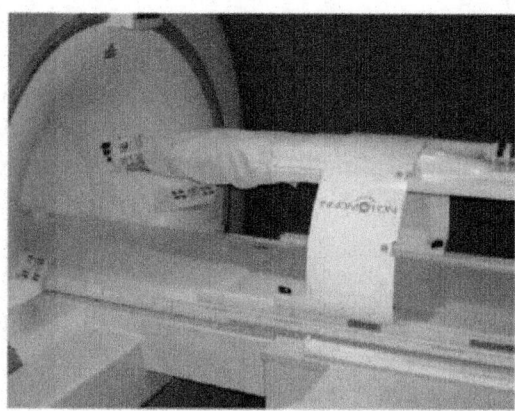

(a) InnoMotion by Synthes Inc.

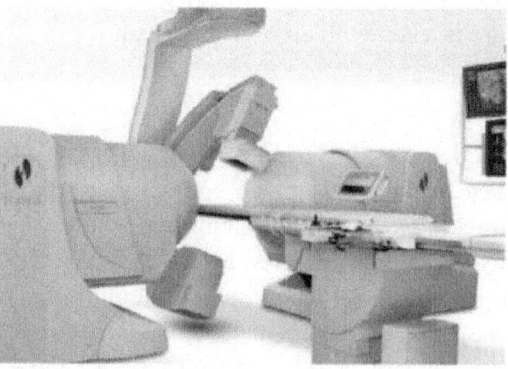

(b) Niobe by Stereotaxis

(c) Sensei X by Hansen Medical

Figure 4: Real-time image guided percutaneous (a) and catheter robots ((b) and (c)). The InnoMotion image is ©2012 Synthes Inc. The Niobe image is ©2012 Stereotaxis. The Sensei X image is ©2012 Hansen Medical.

STEERABLE CATHETERS

Vascular catheterization is used to diagnose and treat various cardiac and vasculature diseases, including direct pressure measurements, biopsy, ablation for atrial fibrillation, and angioplasty for obstructed blood vessels [68–70]. The catheter is inserted into a blood vessel and the portion external to the patient is manipulated to move the catheter tip to the surgical site, while fluoroscopy provides image guidance. Due to the supporting tissue, catheters only require three degrees of freedom, typically: tip flexion, tip rotation, and insertion depth. Possible benefits of robot-steered catheters are shorter procedures, reduced forces exerted on the vasculature by the catheter tip, increased accuracy in catheter positioning, and teleoperation (reducing exposure of the physician to radiation) [71].

The Sensei X (Hansen Medical, FDA clearance and CE mark 2007, previous generation the Sensei, Figure4(c)) uses two steerable sheaths, one inside the other, to create a tight bend radius [72–74]. The sheaths are steered via a remotely operated system of pulleys. IntelliSense force sensing allows constant estimation of the contact forces by gently pulsing the catheter a short distance in and out of the steerable inner sheath and measuring forces at the proximal end of the catheter. These forces are communicated visually as well as through a vibratory feedback to the surgeon's hand on the "3D joystick". Corindus's CorPath 200 is a direct competitor with the Sensei X, but is not yet commercially available.

The Niobe (Stereotaxis, CE mark 2008, FDA clearance 2009) is a remote magnetic navigation system, in which a magnetic field is used to guide the catheter tip [75]. The magnetic field is generated by two permanent magnets contained in housings on either side of a fluoroscopy table (Figure 4(b)). The surgeon manipulates a joystick to specify the desired orientation of the catheter tip, causing the orientations of the magnets to vary under computer-control, and thereby controlling the magnetic field. A second joystick controls advancement/ retraction of the catheter. Chun et al. report significant improvements in surgical outcomes due to advances in the design of magnetically guided catheters [76].

RADIOSURGERY

Radiosurgery is a treatment (not a surgery), in which focused beams of ionizing radiation are directed at the patient, primarily to treat tumors [77, 78]. By directing the beam through the tumor at various orientations, high-dose radiation is delivered to the tumor while the surrounding tissue receives significantly less radiation. Prior to real-time tissue tracking, radiosurgery was practically limited to treating the brain using stereotactic frames mounted to the skull with bone screws. Now that real-time tissue tracking is feasible, systems are commercially available.

The CyberKnife (Accuray Inc., FDA cleared 1999, Figure 5(a)) is a frameless radiosurgery system consisting of a robotic arm holding a linear accelerator, a six degree of freedom robotic patient table called the RoboCouch, and an X-ray imaging system that can take real-time images in two orthogonal orientations simultaneously [79, 80]. The two simultaneous, intraoperative X-ray images are not sufficient to provide good definition of the tumor, but are used to register a high-definition preoperative CT image. The robotic arm can then provide the preplanned radiation dosage with a wide range of orientations. For targets that move during treatment (e.g., due to breathing), the optional synchrony system can optically track the tissue surface, correlate the motion of the tissue surface to the motion of radio-opaque fiducials inserted near the target, and thus continuously predict target motion [81]. The intraoperative tracking obviates the need for a stereotactic frame, reducing patient trauma and making it practical to fractionate the dosage over longer time periods.

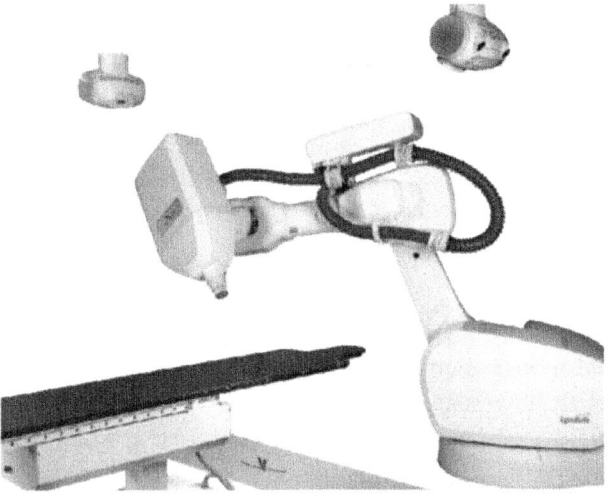

(a) CyberKnife by Accuray Inc.

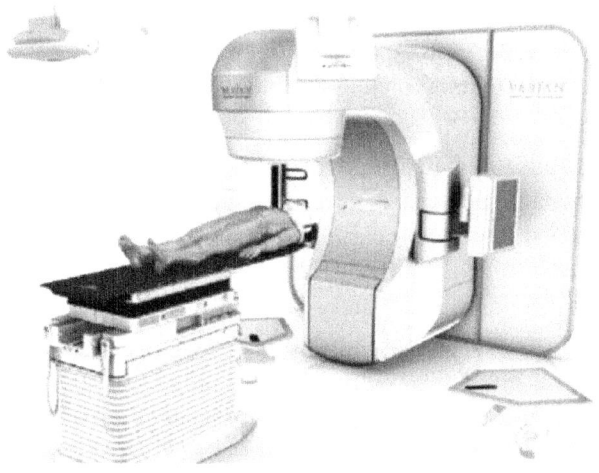

(b) Novalis with TrueBeam STx by BrainLab Inc. and Varian Medical Systems

Figure 5: Radiosurgery robots use X-ray images taken during the treatment to control robotic patient tables, ensuring accurate targeting of the radiosurgery beams. The Cyberknife image and any Accuray trademarks or logos are used with permission from Accuray Incorporated. The image of the Novalis with TrueBeam STx is ©2012 BrianLab Inc.

The Novalis with TrueBeam STx (BrainLab Inc. and Varian Medical Systems, previously Novalis and Trilogy, initial FDA clearance 2000, Figure 5(b)) is also a frameless system with a linear accelerator, but with micro-multileaf collimators for beam shaping [82–84]. Similar to CyberKnife, intraoperative X-rays are compared with a CT, and skin-mounted fiducials are optically tracked in real-time. The delivery system also includes cone beam CT. The patient is moved into position on top of a six degree of freedom robotic couch. The major differences between Cyberknife and Novalis are that the Cyberknife radiation source has more degrees of freedom to be oriented around the patient while the Novalis can shape the radiation beam and claim reduced out-of-field dosage [85, 86].

EMERGENCY RESPONSE

Few medical robot systems are suitable for use outside of the operating room, despite significant research funding on medical devices for disaster response and battlefield medicine. Typical goals for such research include improved extraction of patients from dangerous environments, rapid diagnosis of injuries, and semiautonomous delivery of life-saving interventions. Current Emergency Response robots are little more than single-motor systems, but those systems can be controlled by health monitors to minimize the necessary attention by Emergency Responders. Such a feedback control makes it more likely that such systems will be autonomous, for example, automated external defibrillators.

The AutoPulse Plus (ZOLL Medical Corp., previously by Revivant) is an automated, portable device that combines the functions of the AutoPulse (FDA clearance 2008, Figure 6(a)) cardiopulmonary resuscitation device and the E Series monitor/defibrillator (FDA clearance 2010) [87, 88]. Consisting of a half-backboard containing a battery-powered motor that actuates a chest band, the AutoPulse rhythmically tightens the band to perform chest compressions. The tightness of the band during compressions is a function of the patient's resting chest size, to adjust for interpatient variability. Meanwhile, the E Series monitor/defibrillator measures the rate and depth of chest compressions in real time and filters cardiopulmonary resuscitation artifacts from the electrocardiogram signal. If combined with an automatic battery-powered ventilator, for example, the SAVe (AutoMedx Inc., FDA clearance 2007), basic cardiopulmonary emergency response treatments could be automated while on battery power.

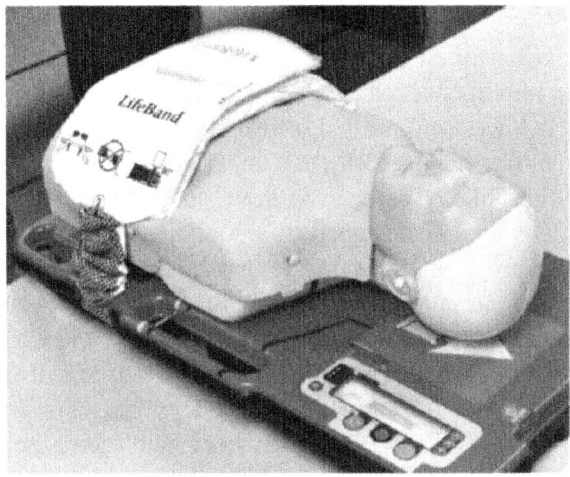

(a) AutoPulse by ZOLL Medical Corp.

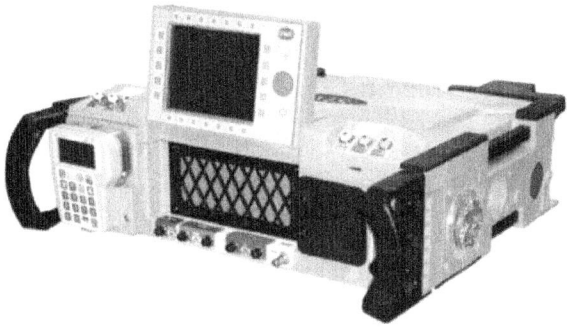

(b) LS-1 by Integrated Medical Systems Inc.

Figure 6: Commercially available Emergency Response robots perform simple actuations compared to robots in other medical disciplines, but their actions are tightly coupled with patient measurements. The AutoPulse image is ©2012 ZOLL Medical Corp. The LS-1 image is ©2012 Integrated Medical Systems Inc.

The LS-1 "suitcase intensive care unit" (Integrated Medical Systems Inc., previous generation called MedEx 1000, previous generation called LSTAT, FDA clearance 2008, Figure 6(b)) takes an inclusive approach to portable life support [89]. The system contains a ventilator with oxygen and carbon dioxide monitoring, electrocardiogram, invasive and noninvasive blood pressure monitoring, fluid/drug infusion pumps, temperature sensing, and blood oxygen level measurement. The LS-1 is battery powered and can be powered by facility or vehicular electrical sources. The system is FDA-cleared for remote control of its diagnostic and therapeutic capabilities.

PROSTHETICS AND EXOSKELETONS

Microprocessor-controlled prosthetics have been available since 1993, specifically the Intelligent Prosthesis knee (Chas. A. Blatchford & Sons, Ltd.). Several microprocessor-controlled prosthetics exist today, predominantly for knee prosthetics, hand prosthetics, and exoskeletons. For example, one current generation knee prosthetic is the C-leg (Otto Bock, FDA clearance 1999, CE mark) which is designed to automatically adjust the swing phase dynamics and improve stability during the stance phase by controlling knee flexion [90]. An example of hand prosthetic is the i-limb ultrahand (Touch Bionics, previous version i-limb hand, FDA clearance and CE mark), the first commercially available hand prosthesis with five individually powered digits, controlled via myoelectric signals generated by muscles in the remaining portion of the patient's limb [91]. For wheelchair users, the ReWalk (Argo Medical Technologies, FDA clearance 2011, CE mark 2010) is one walking assistance exoskeleton that allows users to stand, walk, and climb stairs and is controlled with a wrist-mounted remote and a posture detection sensor [92]. Significant research on exoskeletons is ongoing, such as the research on upper-limb exoskeletons by Rosen and Perry [93]. For further information in the area of prosthetics and exoskeletons see the works by Kazerooni [94] and Bogue [95].

ASSISTIVE AND REHABILITATION SYSTEMS

Assistive robotic systems are designed to allow people with disabilities more autonomy, and they cover a wide range of everyday tasks. In 1992, Handy 1 (Rehab Robotics, Ltd.) became the first commercial assistive robot [96]; it interacts with different trays for tasks such as eating, shaving, and painting, and it is controlled by a single switch input to select the desired action. One task-specific system is the Neater Eater (Neater Solutions Ltd.), a modular device that scoops food from a plate to a person's mouth, and can be controlled manually or via head or foot switches. More general systems rely on arms with many degrees of freedom, such as Exact Dynamics' iARM, a robotic arm with a two-fingered grasper, that attaches to electric wheelchairs and can be controlled via keypad, joystick, or single button.

Rehabilitation systems can be similar to assistive systems, but are designed to facilitate recovery by delivering therapy and measuring the patient's progress, often following a stroke [97]. The Mobility System (Myomo, Inc.) is a wearable robotic device that moves the patient's arm in response to his/her muscle signals, thus creating feedback to facilitate muscle reeducation. The InMotion (Interactive Motion Technologies, based on the MIT-MANUS research platform) is a robotic arm that moves, guides, or perturbs the patient's arm within a planar workspace, while recording motions, velocities, and forces to evaluate progress [98]. For information on research efforts, see Dallaway et al. [99] for an overview of thirty assistive and rehabilitation systems, such as the MASTER II system that uses a rail-mounted robotic arm to make manually controlled, remote-controlled, or preprogrammed motions for various domestic and office tasks [100]. Difficulties in developing rehabilitation robots and potential future uses are investigated by Ceccarelli [101].

Current Research and Development in Medical Robotics

Many more medical robots are currently being researched [16, 19, 21]. Such research will lead to the new capabilities of future commercial systems. This section discusses just a few systems of note.

Raven And Mirosurge

Two prominent academic robot-assisted surgical systems are currently used for research into endoscopic telesurgery: RAVEN II and MiroSurge. The RAVEN II (University of Washington and UC Santa Cruz) is a teleoperated laparoscopic system that was designed to maximize surgical performance based on objective clinical measurements [102–104]. The system has two patient-side arms that are cable-driven with 7 degrees of freedom each. The arm kinematics are based on a spherical mechanism such that the tool always passes through a remote center (e.g., the insertion point for minimally invasive surgery). The length and angles of the links were optimized to maximize performance throughout the workspace. The arms are lighter, smaller, and less expensive than current robotic systems for laparoscopy. The instrument controllers are haptic devices, allowing force feedback on the operator's hands based on tool forces or virtual fixtures (e.g., forbidden regions) defined with respect to patient anatomy (see [105–107] for the impact of haptics on surgery). Teleoperation experiments have been conducted with the RAVEN, including routing the data transmission through an unmanned aircraft. In February 2012, five systems were provided to various other surgical robotics research labs to spur collaboration and further development efforts.

In another endoscopic research effort, the German Aerospace Center (DLR) is developing MiroSurge to be highly versatile with respect to the number of surgical domains, arm-mounting locations, number of robots, different control modes (e.g., control of position versus control of force), and the ability to integrate with other technologies [108]. The expectation is for a base robot system to hold specialized instruments, such as DLR's MICA instrument (which is itself a robotic tool with 3 degrees of freedom and force sensing) [109]. By using a general robotic base to hold a specialized robotic instrument that has its own motors, sensors, and control electronics, the same base system can be specialized for various procedures just by switching the instrument. The base robot, the DLR Miro, masses 10 kg with a 3 kg payload and has serial kinematics that resemble the kinematics of the human arm, with joint ranges and link lengths optimized based on certain medical procedures [110]. Unlike the RAVEN II, the MIRO arm does not have a remote center of motion, and thus must be controlled to direct the instrument through any insertion point, but is more easily able to handle moving insertion points (e.g., through the chest wall during respiration).

Amadeus

Titan Medical Inc. is currently developing Amadeus, a four-armed laparoscopic surgical robot system, to compete with Intuitive Surgical's da Vinci system. The Amadeus uses snakelike multiarticulating arms for improved maneuverability, and the system is being designed to facilitate teleoperation for long-distance surgery. Human trials are planned for late 2013.

NeuroArm and MrBot

At least two renowned research systems are investigating improved MR-compatible robots. The neuroArm (University of Calgary, MacDonald Dettwiler and Associates, IMRIS) is a two-armed, MRI-guided neurosurgical robot actuated via piezoelectric motors [111, 112]. The neuroArm end effectors are equipped with 3 degrees of freedom optical force sensors and are accurate to tens of micrometers. The MrBot (Johns Hopkins University) is a parallel linkage arm designed for MRI-guided access of the prostate gland, actuated by novel pneumatic stepper motors for reduced MR interference [113].

TraumaPod

TraumaPod (highly collaborative, led by SRI International) is a semi-autonomous telerobotic surgical system designed to be rapidly deployable [114]. The surgical cell consists of a surgical robot (da Vinci for Phase I testing), Scrub Nurse Subsystem, Tool Rack System, Supply Dispensing System, Patient Imaging System (a movable X-ray tube), predecessor of the aforementioned LS-1 ("suitcase intensive care unit"), and Supervisory Controller System. The TraumaPod has demonstrated successful teleoperation of a bowel closure and shunt placement on a phantom without a human in the surgical cell. That success implies the potential for increased automation in the operating room, though challenges were reported in sterilization, anesthesia, and robustness.

HeartLander

The heart has long been a target for surgical robots and various systems continue to investigate how best to treat cardiac diseases, particularly while the heart is beating (e.g., see Section 6) [115]. The HeartLander (HeartLander Surgical) is a minimally invasive robot that uses suction to crawl around the surface of the heart [116, 117]. The system is designed for intrapericardial drug delivery, cell transplantation, epicardial atrial ablation, and other such procedures.

Robots In Vivo

Various groups are expanding and exploring the da Vinci system's approach to enhance surgery by increasing the dexterity of the tool inside the patient. One such example is the University of Nebraska's laparoscopic robotic system for research into single-site surgeries [118]. The system has two arms with six degrees of freedom each, and those arms are fully inserted into the abdomen. The expectation is that, by increasing the tool's dexterity inside the patient, fewer incisions will be needed to insert instruments because multiple tools/ arms can pass through a single incision and then spread out inside the patient. Further, miniaturizing the robotics reduces the difficulty in working with (and around) the system in the operating room. A limited number of animal trials (colon resections) have been performed to demonstrate feasibility.

Swallowable capsules take patient trauma reduction to an extreme, but current systems are limited to diagnostic uses. Core temperature measurement has been FDA cleared since 1990, by CorTemp (HQ Inc., formerly HTI Technologies). More recently, capsule endoscopy systems (PillCam by Given Imaging with FDA clearance 2001, and EndoCapsule by Olympus with FDA clearance 2007) consist of a forward-looking wide-angle camera taking regularly timed pictures, a battery, and lights, all contained in a capsule [119, 120]. SmartPill (SmartPill Corp., FDA clearance 2006) utilizes multiple sensors to measure pressure, pH level, gastric emptying time, and bowel emptying time [121]. Sayaka (RF Co Ltd.) is a novel design, not FDA cleared, using a lateral-facing camera that rotates inside the capsule to image the entire tract and is designed without a battery, instead relying on an externally applied magnetic field for inductive power supply [122]. For the future, many enhancements have been proposed, including biopsy, real-time localization of the capsule, drug delivery, ultrasonic imaging, increasing motility by electrically inducing peristalsis, and utilizing an active locomotion system involving treads or legs [123].

In a more dramatic approach to in vivo robotics, micro/nanotechnology is a multibillion dollar area of research [124, 125], including investigation for various medical robotic uses such as inexpensive directable drug delivery vehicles, radio-controlled biomolecules, tissue micromanipulation platforms, artificial mechanical white blood cells, and many other therapeutic approaches that may benefit from robots working at the cellular level [126–129]. Construction of functional systems is an ongoing area of research, particularly with respect to generating and powering motion. Many current prototypes are propelled and guided via magnetic fields, though some utilize external electrical energy sources [130, 131]. To the author's knowledge, clinical trials have not begun for any medical micro/nanorobot.

DISCUSSION

Medical robotics is a young and relatively unexplored field made possible by technical improvements over the past couple of decades. Currently available systems have been available for too short time to allow long-term studies. Nor are the benefits potentially provided by medical robots fully understood. Medical robots have only passed through a few technological generations and the technology continues to change and leap into new areas. Yet by looking at the current market and representative research systems, educated guesses can be made about the impacts of robots on near-future medicine.

In surgical robotics, there has been a trend away from autonomous or even semiautonomous motions, and toward synergistic manipulation and virtual fixtures. Thus, the robot acts as a guidance tool, providing information (and possibly a physical nudge) to keep the surgeon on target. Such use requires accurate localization of the tissues in the surgical site, even as the tissues are manipulated during surgery. Improved imaging systems (e.g., Explorer, an intraoperative soft tissue tracker by Pathfinder Therapeutics [132]) or robot compatibility with MRI or CT will provide that localization. In particular, MRI-guided robots will benefit from intraoperative 3D images with excellent soft tissue contrast and accurate registration between the tool and the tissue, thus allowing precise virtual fixtures, "snap-to" and "stand-off" behaviors. Further, such imaging will allow modeling and rapid prototyping of patient-specific templates/jigs/implants.

The physical designs for medical robots will continue to improve, reducing expense and size, while minimizing or compensating for nonidealities such as flexion, for example, the CRIGOS robot [133]. With better physical designs, semiautonomous behavior will likely become more useful. "Macros" may become commonplace, wherein the surgeon presses a button and the robot performs a preprogrammed motion, such as passing a suture needle between graspers, or the Sensei's autoretract feature [13].

Robots will see more use for medical training purposes, bolstered by improved tissue-modeling capabilities, by the increasing objectivity in healthcare assessment, by advances in computer simulations, and as a result of increased data mining arising naturally from improved data connectivity between devices and between institutions. Some such systems are already available, such as the aforementioned da Vinci Skills Simulator, the Virtual I.V. Simulator by Laerdal, and the EndoscopyVR Surgical Simulator by CAE. For the same reasons, robotics will continue to make possible new medical procedures and treatments, such as new Single-Port Access procedures.

Even as robots are developed for new medical areas, other tools may encroach on medical needs currently filled by robots. Medical robots must develop a firm basis in improved medical outcomes, or risk being displaced by pharmaceuticals, tissue engineering, gene therapy, and rapid innovation in manual tools (e.g., the SPIDER Surgical System by TransEnterix, and the EndoStitch by Covidien). To that end, improvements in medical robotics must address and solve real problems in healthcare, ultimately providing a clear improvement in quality of life when compared with the alternatives.

REFERENCES

1. Intuitive Surgical Incorporation, "Webpage on da Vinci clinical evidence," March 2012,http://www.intuitivesurgical.com/company/clinical-evidence.

2. Intuitive Surgical Incorporation, "Webpage on da Vinci regulatory approval," March 2012,http://www.intuitivesurgical.com/company/regulatory-clearance.html.

3. H. Lavery, D. Samadi, and A. Levinson, "Not a zero-sum game: the adoption of robotics has increased overall prostatectomy utilization in the united states," in Proceedings of the American Urological Association Annual Meeting, Poster Session, Washington, DC, USA, 2011.

4. R. D. Howe and Y. Matsuoka, "Robotics for surgery," Annual Review of Biomedical Engineering, vol. 1, pp. 211–240, 1999. ·

5. B. Davies, "A review of robotics in surgery," Proceedings of the Institution of Mechanical Engineers H, vol. 214, no. 1, pp. 129–140, 2000.

6. R. H. Taylor and D. Stoianovici, "Medical robotics in computer-integrated surgery," IEEE Transactions on Robotics and Automation, vol. 19, no. 5, pp. 765–781, 2003.

7. R. Lanfranco, A. E. Castellanos, J. P. Desai, and W. C. Meyers, "Robotic surgery: a current perspective," Annals of Surgery, vol. 239, no. 1, pp. 14–21, 2004.

8.   P. Berkelman, J. Troccaz, and P. Cinquin, "Body-supported medical robots: a survey," Journal of Robotics and Mechatronics, vol. 16, pp. 513–519, 2004.

9.   L. Guo, X. Pan, Q. Li, F. Zheng, and Z. Bao, "A survey on the gastrointestinal capsule micro-robot based on wireless and optoelectronic technology," Journal of Nanoelectronics and Optoelectronics, vol. 7, no. 2, pp. 123–127, 2012.

10.   C. Stüer, F. Ringel, M. Stoffel, A. Reinke, M. Behr, and B. Meyer, "Robotic technology in spine surgery: current applications and future developments," Intraoperative Imaging, vol. 109, pp. 241–245, 2011.·

11.   S. Badaan and D. Stoianovici, "Robotic systems: past, present, and future," in Robotics in Genitourinary Surgery, pp. 655–665, Springer, New York, NY, USA, 2011.

12.   Singh, "Robotics in urological surgery: review of current status and maneuverability, and comparison of robot-assisted and traditional laparoscopy," Computer Aided Surgery, vol. 16, no. 1, pp. 38–45, 2011.

13.   G. P. Moustris, S. C. Hiridis, K. M. Deliparaschos, and K. M. Konstantinidis, "Evolution of autonomous and semi-autonomous robotic surgical systems: a review of the literature," International Journal of Medical Robotics and Computer Assisted Surgery, vol. 7, no. 4, pp. 375–392, 2011.

14.   B. Challacombe and D. Stoianovici, "The basic science of robotic surgery," in Urologic Robotic Surgery in Clinical Practice, pp. 1–23, 2009.

15.   H. Kenngott, L. Fischer, F. Nickel, J. Rom, J. Rassweiler, and B. Muller-Stich, "Status of robotic assistance: a less traumatic and more accurate minimally invasive surgery?" Langenbeck's Archives of Surgery, vol. 397, no. 3, pp. 1–9, 2012.

16.   P. Gomes, "Surgical robotics: reviewing the past, analysing the present, imagining the future," Robotics and Computer-Integrated Manufacturing, vol. 27, no. 2, pp. 261–266, 2011.

17.   M. Okamura, M. J. Matarić, and H. I. Christensen, "Medical and health-care robotics," IEEE Robotics and Automation Magazine, vol. 17, no. 3, pp. 26–37, 2010.

18.   S. Najarian, M. Fallahnezhad, and E. Afshari, "Advances in medical robotic systems with specific applications in surgery—a review," Journal of Medical Engineering and Technology, vol. 35, no. 1, pp. 19–33, 2011.

19.   Rosen, B. Hannaford, and R. Satava, Eds., Surgical Robotics: Systems

Applications and Visions, Springer, New York, NY, USA, 2011.

20. N. Nathoo, M. C. Çavuşo☐lu, M. A. Vogelbaum, and G. H. Barnett, "In touch with robotics: neurosurgery for the future," Neurosurgery, vol. 56, no. 3, pp. 421–431, 2005.

21. T. Haidegger, L. Kovacs, G. Fordos, Z. Benyo, and P. Kazanzides, "Future trends in robotic neurosurgery," in Proceedings of the 14th Nordic-Baltic Conference on Biomedical Engineering and Medical Physics (NBC '08), pp. 229–233, Springer, June 2008.

22. Y. S. Kwoh, J. Hou, E. A. Jonckheere, and S. Hayati, "A robot with improved absolute positioning accuracy for CT guided stereotactic brain surgery," IEEE Transactions on Biomedical Engineering, vol. 35, no. 2, pp. 153–160, 1988.

23. D. Glauser, H. Fankhauser, M. Epitaux, J. L. Hefti, and A. Jaccottet, "Neurosurgical robot Minerva: first results and current developments," Journal of Image Guided Surgery, vol. 1, no. 5, pp. 266–272, 1995.·

24. T. R. K. Varma and P. Eldridge, "Use of the NeuroMate stereotactic robot in a frameless mode for functional neurosurgery," International Journal of Medical Robotics and Computer Assisted Surgery, vol. 2, no. 2, pp. 107–113, 2006.

25. Q. H. Li, L. Zamorano, A. Pandya, R. Perez, J. Gong, and F. Diaz, "The application accuracy of the NeuroMate robot—a quantitative comparison with frameless and frame-based surgical localization systems," Computer Aided Surgery, vol. 7, no. 2, pp. 90–98, 2002.

26. P. Morgan, T. Carter, S. Davis et al., "The application accuracy of the pathfinder neurosurgical robot," in International Congress Series, vol. 1256, pp. 561–567, Elsevier, Amsterdam, The Netherlands, 2003.

27. G. Deacon, A. Harwood, J. Holdback et al., "The pathfinder image-guided surgical robot," Proceedings of the Institution of Mechanical Engineers H, vol. 224, no. 5, pp. 691–713, 2010.

28. Brodie and S. Eljamel, "Evaluation of a neurosurgical robotic system to make accurate burr holes,"International Journal of Medical Robotics and Computer Assisted Surgery, vol. 7, no. 1, pp. 101–106, 2011.

29. Joskowicz, R. Shamir, Z. Israel, Y. Shoshan, and M. Shoham, "Renaissance robotic system for keyhole cranial neurosurgery: in-vitro accuracy study," in Proceedings of the Simposio Mexicano en Ciruga Asistida por Computadora y Procesamiento de Imgenes Mdicas (MexCAS '11), 2011.

30. D. P. Devito, L. Kaplan, R. Dietl et al., "Clinical acceptance and accuracy

assessment of spinal implants guided with spineassist surgical robot: retrospective study," Spine, vol. 35, no. 24, pp. 2109–2115, 2010.

31. Yang, J. Jung, J. Kim et al., "Current and future of spinal robot surgery," Korean Journal of Spine, vol. 7, no. 2, pp. 61–65, 2010.

32. J. E. Lang, S. Mannava, A. J. Floyd et al., "Robotic systems in orthopaedic surgery," Journal of Bone and Joint Surgery B, vol. 93, no. 10, pp. 1296–1299, 2011.

33. W. L. Bargar, A. Bauer, and M. Börner, "Primary and revision total hip replacement using the robodoc system," Clinical Orthopaedics and Related Research, vol. 354, pp. 82–91, 1998. ·

34. P. Schulz, K. Seide, C. Queitsch et al., "Results of total hip replacement using the Robodoc surgical assistant system: clinical outcome and evaluation of complications for 97 procedures," International Journal of Medical Robotics and Computer Assisted Surgery, vol. 3, no. 4, pp. 301–306, 2007.

35. P. Kazanzides, J. Zuhars, B. Mittelstadt, and R. H. Taylor, "Force sensing and control for a surgical robot," in Proceedings of the IEEE International Conference on Robotics and Automation, pp. 612–617, May 1992.

36. D. Pearle, P. F. O'Loughlin, and D. O. Kendoff, "Robot-assisted unicompartmental knee arthroplasty," Journal of Arthroplasty, vol. 25, no. 2, pp. 230–237, 2010.

37. D. Pearle, D. Kendoff, V. Stueber, V. Musahl, and J. A. Repicci, "Perioperative management of unicompartmental knee arthroplasty using the MAKO robotic arm system (MAKOplasty)," American Journal of Orthopedics, vol. 38, no. 2, pp. 16–19, 2009. ·

38. Plaskos, P. Cinquin, S. Lavallée, and A. J. Hodgson, "Praxiteles: a miniature bone-mounted robot for minimal access total knee arthroplasty," The International Journal of Medical Robotics and Computer Assisted Surgery, vol. 1, no. 4, pp. 67–79, 2005.

39. Koulalis, P. F. O'Loughlin, C. Plaskos, D. Kendoff, M. B. Cross, and A. D. Pearle, "Sequential versus automated cutting guides in computer-assisted total knee arthroplasty," Knee, vol. 18, no. 6, pp. 436–442, 2010.

40. G. Brisson, T. Kanade, A. DiGioia, and B. Jaramaz, "Precision freehand sculpting of bone," inProceedings of the 7th International Conference on Medical Image Computing and Computer-Assisted Intervention (MICCAI '04), pp. 105–112, September 2004.

41. G. Brisson, The Precision Freehand Sculptor: a Robotic Tool for Less Invasive Joint Replacement Surgery, ProQuest, 2008.

42. P. L. Yen and B. L. Davies, "Active constraint control for image-guided robotic surgery," Proceedings of the Institution of Mechanical Engineers H, vol. 224, no. 5, pp. 623–631, 2010.

43. G. Harrell and B. T. Heniford, "Minimally invasive abdominal surgery: lux et veritas past, present, and future," American Journal of Surgery, vol. 190, no. 2, pp. 239–243, 2005.

44. G. Dogangil, B. L. Davies, and F. Rodriguez Y Baena, "A review of medical robotics for minimally invasive soft tissue surgery," Proceedings of the Institution of Mechanical Engineers H, vol. 224, no. 5, pp. 653–679, 2010.

45. Kuo and J. Dai, "Robotics for minimally invasive surgery: a historical review from the perspective of kinematics," in Proceedings of the International Symposium on History of Machines and Mechanisms, pp. 337–354, Springer, 2009.

46. S. J. Harris, F. Arambula-Cosio, and Q. Mei, "The probot—an active robot for prostate resection,"Proceedings of the Institution of Mechanical Engineers H, vol. 211, no. 4, pp. 317–325, 1997. ·

47. G. H. Ballantyne, "Robotic surgery, telerobotic surgery, telepresence, and telementoring: review of early clinical results," Surgical Endoscopy and Other Interventional Techniques, vol. 16, no. 10, pp. 1389–1402, 2002.

48. G. T. Sung and I. S. Gill, "Robotic laparoscopic surgery: a comparison of the da Vinci and Zeus systems," Urology, vol. 58, no. 6, pp. 893–898, 2001.

49. J. Marescaux, J. Leroy, M. Gagner et al., "Transatlantic robot-assisted telesurgery," Nature, vol. 413, no. 6854, pp. 379–380, 2001.

50. P. Mozer, J. Troccaz, and D. Stoinaovici, "Robotics in urology: past, present, and future," in Atlas of Robotic Urologic Surgery, L. Su, Ed., Current Clinical Urology, ch. 1, pp. 3–13, Springer, New York, NY, USA, 2011.

51. K. Shah and R. Abaza, "Comparison of intraoperative outcomes using the new and old generation da Vinci robot for robot-assisted laparoscopic prostatectomy," British Journal of Urology International, vol. 108, no. 10, pp. 1642–1645, 2011.

52. J. Bodner, H. Wykypiel, G. Wetscher, and T. Schmid, "First experiences with the da Vinci operating robot in thoracic surgery," European Journal of Cardio-Thoracic Surgery, vol. 25, no. 5, pp. 844–851, 2004.

53. Tewari, A. Srivasatava, and M. Menon, "A prospective comparison of radical retropubic and robot-assisted prostatectomy: experience in one

institution," British Journal of Urology International, vol. 92, no. 3, pp. 205–210, 2003.

54. S. Maeso, M. Reza, J. A. Mayol et al., "Efficacy of the da Vinci surgical system in abdominal surgery compared with that of laparoscopy: a systematic review and meta-analysis," Annals of Surgery, vol. 252, no. 2, pp. 254–262, 2010.

55. R. E. Link, S. B. Bhayani, and L. R. Kavoussi, "A prospective comparison of robotic and laparoscopic pyeloplasty," Annals of Surgery, vol. 243, no. 4, pp. 486–491, 2006.

56. Amodeo, A. Linares Quevedo, J. V. Joseph, E. Belgrano, and H. R. H. Patel, "Robotic laparoscopic surgery: cost and training," Minerva Urologica e Nefrologica, vol. 61, no. 2, pp. 121–128, 2009. ·

57. W. Jeong, F. Petros, and C. Rogers, Robotic Surgery: Basic Instrumentation and Troubleshooting, ch. 72, Wiley-Blackwell, Hoboken, NJ, USA, 2012.

58. A. Lerner, M. Ayalew, W. J. Peine, and C. P. Sundaram, "Does training on a virtual reality robotic simulator improve performance on the da Vinci surgical system?" Journal of Endourology, vol. 24, no. 3, pp. 467–472, 2010.

59. K. Cleary and T. M. Peters, "Image-guided interventions: technology review and clinical applications,"Annual Review of Biomedical Engineering, vol. 12, pp. 119–142, 2010.

60. M. E. Hagen, O. J. Wagner, I. Inan et al., "Robotic single-incision transabdominal and transvaginal surgery: initial experience with intersecting robotic arms," International Journal of Medical Robotics and Computer Assisted Surgery, vol. 6, no. 3, pp. 251–255, 2010.

61. M. Kroh, K. El-Hayek, S. Rosenblatt et al., "First human surgery with a novel single-port robotic system: cholecystectomy using the da Vinci Single-Site platform," Surgical Endoscopy, vol. 25, no. 11, pp. 3566–3573, 2011.

62. M. Stark, T. Benhidjeb, S. Gidaro, and E. Morales, "The future of telesurgery: a universal system with haptic sensation," Journal of the Turkish-German Gynecological Association, vol. 13, no. 1, pp. 74–76, 2012.

63. S. DiMaio and S. Salcudean, "Needle steering and model-based trajectory planning," in Proceedings of the 6th International Conference on Medical Image Computing and Computer-Assisted Intervention (MICCAI '03), pp. 33–40, 2003.

64. H. Delingette, "Toward realistic soft-tissue modeling in medical

simulation," Proceedings of the IEEE, vol. 86, no. 3, pp. 512–523, 1998. ·

65. Melzer, B. Gutmann, T. Remmele et al., "Innomotion for percutaneous image-guided interventions,"IEEE Engineering in Medicine and Biology Magazine, vol. 27, no. 3, pp. 66–73, 2008.

66. M. Li, A. Kapoor, D. Mazilu, and K. A. Horvath, "Pneumatic actuated robotic assistant system for aortic valve replacement under MRI guidance," IEEE Transactions on Biomedical Engineering, vol. 58, no. 2, pp. 443–451, 2011.

67. S. Zangos, A. Melzer, K. Eichler et al., "MR-compatible assistance system for biopsy in a high-field-strength system: initial results in patients with suspicious prostate lesions," Radiology, vol. 259, no. 3, pp. 903–910, 2011.

68. H. J. Swan, W. Ganz, J. Forrester, H. Marcus, G. Diamond, and D. Chonette, "Catheterization of the heart in man with use of a flow-directed balloon-tipped catheter," The New England Journal of Medicine, vol. 283, no. 9, pp. 447–451, 1970. ·

69. M. R. Franz, D. Burkhoff, and H. Spurgeon, "In vitro validation of a new cardiac catheter technique for recording monophasic action potentials," European Heart Journal, vol. 7, no. 1, pp. 34–41, 1986. ·

70. J. M. Gore, R. J. Goldberg, D. H. Spodick, J. S. Alpert, and J. E. Dalen, "A community-wide assessment of the use of pulmonary artery catheters in patients with acute myocardial infarction," Chest, vol. 92, no. 4, pp. 721–727, 1987. ·

71. Steven, H. Servatius, T. Rostock et al., "Reduced fluoroscopy during atrial fibrillation ablation: benefits of robotic guided navigation," Journal of Cardiovascular Electrophysiology, vol. 21, no. 1, pp. 6–12, 2010.

72. V. Y. Reddy, P. Neuzil, Z. J. Malchano et al., "View-synchronized robotic image-guided therapy for atrial fibrillation ablation: experimental validation and clinical feasibility," Circulation, vol. 115, no. 21, pp. 2705–2714, 2007.

73. K. R. J. Chun, B. Schmidt, B. Köktürk et al., "Catheter ablation—new developments in robotics," Herz, vol. 33, no. 8, pp. 586–589, 2008.

74. C. V. Riga, C. D. Bicknell, D. Wallace, M. Hamady, and N. Cheshire, "Robot-assisted antegrade in-situ fenestrated stent grafting," CardioVascular and Interventional Radiology, vol. 32, no. 3, pp. 522–524, 2009.

75. S. Ernst, F. Ouyang, C. Linder et al., "Initial experience with remote catheter ablation using a novel magnetic navigation system," Circulation, vol. 109, no. 12, pp. 1472–1475, 2004.

76. J. K. R. Chun, S. Ernst, S. Matthews et al., "Remote-controlled catheter ablation of accessory pathways: results from the magnetic laboratory," European Heart Journal, vol. 28, no. 2, pp. 190–195, 2007.

77. L. Leksell, "Stereotactic radiosurgery," Journal of Neurology Neurosurgery and Psychiatry, vol. 46, no. 9, pp. 797–803, 1983.

78. R. Schulz and N. Agazaryan, Shaped-Beam Radiosurgery: State of the Art, Springer, New York, NY, USA, 2011.

79. J. R. Adler Jr., S. D. Chang, M. J. Murphy, J. Doty, P. Geis, and S. L. Hancock, "The cyberknife: a frameless robotic system for radiosurgery," Stereotactic and Functional Neurosurgery, vol. 69, no. 1–4, pp. 124–128, 1997.

80. J. Gagnon, N. M. Nasr, J. J. Liao et al., "Treatment of spinal tumors using cyberKnife fractionated stereotactic radiosurgery: pain and quality-of-life assessment after treatment in 200 patients,"Neurosurgery, vol. 64, no. 2, pp. 297–306, 2009.

81. M. Hoogeman, J. B. Prevost, J. Nuyttens, J. Poll, P. Levendag, and B. Heijmen, "Clinical accuracy of the respiratory tumor tracking system of the cyberknife: assessment by analysis of log files," International Journal of Radiation Oncology *Biology* Physics, vol. 74, no. 1, pp. 297–303, 2009.

82. J. P. Rock, S. Ryu, F. F. Yin, F. Schreiber, and M. Abdulhak, "The evolving role of stereotactic radiosurgery and stereotactic radiation therapy for patients with spine tumors," Journal of Neuro-Oncology, vol. 69, no. 1–3, pp. 319–334, 2004.

83. R. E. Wurm, S. Erbel, I. Schwenkert et al., "Novalis frameless image-guided noninvasive radiosurgery: initial experience," Neurosurgery, vol. 62, no. 5, pp. A11–A17, 2008. ·

84. Z. Chang, T. Liu, J. Cai, Q. Chen, Z. Wang, and F. Yin, "Evaluation of integrated respiratory gating systems on a novalis tx system," Journal of Applied Clinical Medical Physics, vol. 12, no. 3, article 3495, 2011.

85. Liu, N. Agazaryan, C. Yu, H. Han, T. Schultheiss, and J. Wong, "A multi-center consortium study of competing platforms for intracranial stereotactic irradiation," International Journal of Radiation Oncology *Biology* Physics, vol. 72, supplement 1, pp. S213–S213, 2008.

86. M. Abacioglu, "Advances in technology in radiation oncology," Oncology, vol. 2, no. 1, pp. 11–14, 2012.

87. R. Halperin, N. Paradis, J. P. Ornato et al., "Cardiopulmonary resuscitation with a novel chest compression device in a porcine model of

cardiac arrest: improved hemodynamics and mechanisms,"Journal of the American College of Cardiology, vol. 44, no. 11, pp. 2214–2220, 2004.

88.   Hallstrom, T. D. Rea, M. R. Sayre et al., "Manual chest compression vs use of an automated chest compression device during resuscitation following out-of-hospital cardiac arrest: a randomized trial,"Journal of the American Medical Association, vol. 295, no. 22, pp. 2620–2628, 2006.

89.   R. Palmer, "Integrated diagnostic and treatment devices for enroute critical care of patients within theater," in Proceedings of the RTO Human Factors and Medicine Panel Symposium, Amsterdam, The Netherlands, October 2010.

90.   R. Seymour, B. Engbretson, K. Kott et al., "Comparison between the C-leg microprocessor-controlled prosthetic knee and non-microprocessor control prosthetic knees: a preliminary study of energy expenditure, obstacle course performance, and quality of life survey," Prosthetics and Orthotics International, vol. 31, no. 1, pp. 51–61, 2007.

91.   Otr, H. A. Reinders-Messelink, R. M. Bongers, H. Bouwsema, and C. K. Van Der Sluis, "The i-LIMB hand and the DMC plus hand compared: a case report," Prosthetics and Orthotics International, vol. 34, no. 2, pp. 216–220, 2010.

92.   K. Low, "Robot-assisted gait rehabilitation: from exoskeletons to gait systems," in Proceedings of the Defense Science Research Conference and Expo (DSR '11), pp. 1–10, August 2011.

93.   J. Rosen and J. C. Perry, "Upper limb powered exoskeleton," International Journal of Humanoid Robotics, vol. 4, no. 3, pp. 529–548, 2007.

94.   Kazerooni, "Exoskeletons for human performance augmentation," in Springer Handbook of Robotics, B. Siciliano and O. Khatib, Eds., Springer, New York, NY, USA, 2008.

95.   R. Bogue, "Exoskeletons and robotic prosthetics: a review of recent developments," Industrial Robot, vol. 36, no. 5, pp. 421–427, 2009.

96.   M. J. Topping and J. K. Smith, "The development of Handy 1. A robotic system to assist the severely disabled," Technology and Disability, vol. 10, no. 2, pp. 95–105, 1999. ·

97.   M. Hillman, "Rehabilitation robotics from past to present—a historical perspective," in Advances in Rehabilitation Robotics, pp. 25–44, Springer, New York, NY, USA, 2004.

98.   Waldner, C. Werner, and S. Hesse, "Robot assisted therapy in neurorehabilitation," Europa Medicophysica, vol. 44, supplement 1, pp.

1–3, 2008.

99. L. Dallaway, R. D. Jackson, and P. H. A. Timmers, "Rehabilitation robotics in Europe," IEEE Transactions on Rehabilitation Engineering, vol. 3, no. 1, pp. 35–45, 1995.

100. M. Busnel, R. Cammoun, F. Coulon-Lauture, J. M. Détriché, G. Le Claire, and B. Lesigne, "The robotized workstation "MASTER" for users with tetraplegia: description and evaluation," Journal of Rehabilitation Research and Development, vol. 36, no. 3, pp. 217–229, 1999. ·

101. M. Ceccarelli, "Problems and issues for service robots in new applications," International Journal of Social Robotics, vol. 3, no. 3, pp. 299–312, 2011.

102. M. J. H. Lum, D. C. W. Friedman, G. Sankaranarayanan et al., "The RAVEN: design and validation of a telesurgery system," International Journal of Robotics Research, vol. 28, no. 9, pp. 1183–1197, 2009.

103. Simorov, R. Otte, C. Kopietz, and D. Oleynikov, "Review of surgical robotics user interface: what is the best way to control robotic surgery?" Surgical Endoscopy, vol. 26, no. 8, pp. 2117–2125, 2012.

104. Kuo, J. Dai, and P. Dasgupta, "Kinematic design considerations for minimally invasive surgical robots: an overview," The International Journal of Medical Robotics and Computer Assisted Surgery, vol. 8, no. 2, pp. 127–145, 2012.

105. Wagner, N. Stylopoulos, and R. Howe, "The role of force feedback in surgery: analysis of blunt dissection," in Proceedings of the 10th Symposium on Haptic Interfaces for Virtual Environment and Teleoperator Systems, vol. 2002, Citeseer, 2002.

106. M. Tavakoli, R. V. Patel, and M. Moallem, "Haptic interaction in robot-assisted endoscopic surgery: a sensorized end-effector," The International Journal of Medical Robotics and Computer Assisted Surgery, vol. 1, no. 2, pp. 53–63, 2005.

107. M. Okamura, "Haptic feedback in robot-assisted minimally invasive surgery," Current Opinion in Urology, vol. 19, no. 1, pp. 102–107, 2009.

108. U. Hagn, R. Konietschke, A. Tobergte et al., "DLR MiroSurge: a versatile system for research in endoscopic telesurgery," International Journal of Computer Assisted Radiology and Surgery, vol. 5, no. 2, pp. 183–193, 2010.

109. S. Thielmann, U. Seibold, R. Haslinger et al., "MICA—a new generation of versatile instruments in robotic surgery," in Proceedings of the 23rd IEEE/RSJ International Conference on Intelligent Robots and Systems

(IROS '10), pp. 871–878, October 2010.

110. R. Konietschke, T. Ortmaier, H. Weiss, G. Hirzinger, and R. Engelke, "Manipulability and accuracy measures for a medical robot in minimally invasive surgery," in Advances in Robot Kinematics, 2004.

111. G. Sutherland, P. McBeth, and D. Louw, "Neuroarm: an mr compatible robot for microsurgery," inInternational Congress Series, vol. 1256, pp. 504–508, Elsevier, Amsterdam, The Netherland, 2003.

112. M. J. Lang, A. D. Greer, and G. R. Sutherland, "Intra-operative robotics: NeuroArm," Intraoperative Imaging, vol. 109, pp. 231–236, 2011. ·

113. D. Stoianovici, D. Song, D. Petrisor et al., "'MRI Stealth' robot for prostate interventions," Minimally Invasive Therapy and Allied Technologies, vol. 16, no. 4, pp. 241–248, 2007.

114. Garcia, J. Rosen, C. Kapoor et al., "Trauma pod: a semi-automated telerobotic surgical system,"International Journal of Medical Robotics and Computer Assisted Surgery, vol. 5, no. 2, pp. 136–146, 2009.

115. S. G. Yuen, P. M. Novotny, and R. D. Howe, "Quasiperiodic predictive filtering for robot-assisted beating heart surgery," in Proceedings of the IEEE International Conference on Robotics and Automation (ICRA '08), pp. 3875–3880, May 2008.

116. N. Patronik, C. Riviere, S. El Qarra, and M. Zenati, "The heartlander: a novel epicardial crawling robot for myocardial injections," in International Congress Series, vol. 1281, pp. 735–739, Elsevier, Amsterdam, The Netherland, 2005.

117. D. Moral Del Agua, N. A. Wood, and C. N. Riviere, "Improved synchronization of heartlander locomotion with physiological cycles," in Proceedings of the 37th Annual Northeast Bioengineering Conference (NEBEC '11), April 2011.

118. T. Wortman, A. Meyer, O. Dolghi et al., "Miniature surgical robot for laparoendoscopic single-incision colectomy," Surgical Endoscopy, vol. 26, pp. 727–731, 2012.

119. Y. Hayashi, H. Yamamoto, T. Yano, and K. Sugano, "Review: diagnosis and management of mid-gastrointestinal bleeding by double-balloon endoscopy," Therapeutic Advances in Gastroenterology, vol. 2, no. 2, pp. 109–117, 2009.

120. Van Gossum, M. M. Navas, I. Fernandez-Urien et al., "Capsule endoscopy versus colonoscopy for the detection of polyps and cancer," The New England Journal of Medicine, vol. 361, no. 3, pp. 264–270, 2009.

121. D. Cassilly, S. Kantor, L. C. Knight et al., "Gastric emptying of a

non-digestible solid: assessment with simultaneous SmartPill pH and pressure capsule, antroduodenal manometry, gastric emptying scintigraphy," Neurogastroenterology and Motility, vol. 20, no. 4, pp. 311–319, 2008.

122. C. Mc Caffrey, O. Chevalerias, C. O'Mathuna, and K. Twomey, "Swallowable-capsule technology," IEEE Pervasive Computing, vol. 7, no. 1, pp. 23–29, 2008.

123. Moglia, A. Menciassi, and P. Dario, "Recent patents on wireless capsule endoscopy," Recent Patents on Biomedical Engineering, vol. 1, no. 1, pp. 24–33, 2008.

124. M. C. Roco, "Nanotechnology: convergence with modern biology and medicine," Current Opinion in Biotechnology, vol. 14, no. 3, pp. 337–346, 2003.

125. M. Copot, A. Popescu, I. Lung, and A. Moldovanu, "Achievements and perspectives in the field of nanorobotics," The Romanian Review Precision Mechanics, Optics and Mechatronics, vol. 19, no. 36, pp. 61–66, 2009.

126. A. Freitas, "What is nanomedicine?" Nanomedicine: Nanotechnology, Biology, and Medicine, vol. 1, no. 1, pp. 2–9, 2005.

127. L. Zhang, J. J. Abbott, L. Dong, B. E. Kratochvil, D. Bell, and B. J. Nelson, "Artificial bacterial flagella: fabrication and magnetic control," Applied Physics Letters, vol. 94, no. 6, Article ID 064107, 2009.

128. G. Kósa, M. Shoham, and M. Zaaroor, "Propulsion of a swimming micro medical robot," in Proceedings of the IEEE International Conference on Robotics and Automation (ICRA '05), pp. 1327–1331, April 2005.

129. G. Dogangil, O. Ergeneman, J. J. Abbott et al., "Toward targeted retinal drug delivery with wireless magnetic microrobots," in Proceedings of the IEEE/RSJ International Conference on Intelligent Robots and Systems (IROS '08), pp. 1921–1926, September 2008.

130. H. Li, J. Tan, and M. Zhang, "Dynamics modeling and analysis of a swimming microrobot for controlled drug delivery," IEEE Transactions on Automation Science and Engineering, vol. 6, no. 2, pp. 220–227, 2009.

131. T. Ebefors, J. Mattsson, E. Kalvesten, and G. Stemme, "A walking silicon micro-robot," in Proceedings of the 10th International Conference on Solid-State Sensors and Actuators (Transducers '99), pp. 1202–1205, 1999.

132. P. Dumpuri, L. W. Clements, B. M. Dawant, and M. I. Miga, "Model-updated image-guided liver surgery: preliminary results using surface characterization," Progress in Biophysics and Molecular Biology, vol. 103, no. 2-3, pp. 197–207, 2010.

133. G. Brandt, A. Zimolong, L. Carrât et al., "CRIGOS: a compact robot for image-guided orthopedic surgery," IEEE Transactions on Information Technology in Biomedicine, vol. 3, no. 4, pp. 252–260, 1999.·

Chapter 10

NEW TRENDS IN ROBOTICS FOR AGRICULTURE: INTEGRATION AND ASSESSMENT OF A REAL FLEET OF ROBOTS

Luis Emmi,[1] Mariano Gonzalez-de-Soto,[1] Gonzalo Pajares,[2] and Pablo Gonzalez-de-Santos[1]

[1]Centre for Automation and Robotics (UPM-CSIC), Arganda del Rey, 28500 Madrid, Spain

[2]Department of Software Engineering and Artificial Intelligence, Faculty of Informatics, University Complutense of Madrid, 28040 Madrid, Spain

ABSTRACT

Computer-based sensors and actuators such as global positioning systems, machine vision, and laser-based sensors have progressively been incorporated into mobile robots with the aim of configuring autonomous systems capable of shifting operator activities in agricultural tasks. However, the incorporation of many electronic systems into a robot impairs its reliability and increases its cost. Hardware minimization, as well as software minimization and ease of integration, is essential to obtain feasible robotic systems. A step forward in the application of automatic equipment in agriculture is the use of fleets of robots, in which a number of specialized robots collaborate to accomplish one or several agricultural tasks. This paper strives to develop a system architecture for both individual robots and robots working in fleets to improve reliability, decrease complexity and costs, and permit the integration of software from different developers. Several solutions are studied, from a fully distributed to a whole integrated architecture in which a central computer runs all processes. This work also studies diverse topologies for controlling fleets of robots and advances other prospective topologies. The architecture presented in this paper is being successfully applied in the RHEA fleet, which comprises three ground mobile units based on a commercial tractor chassis.

INTRODUCTION

In the last twenty years, specialized sensors (machine vision, global positioning systems (GPS) real-time kinematics (RTK), laser-based equipment, and inertial devices), actuators (hydraulic cylinder, linear, and rotational electrical motors), and electronic equipment (embedded computers, industrial PC, and PLC) have enabled the integration of many autonomous vehicles, particularly agricultural robots [1–5]. These autonomous/semiautonomous systems provide accurate positioning and guidance in the working field, which makes them capable of conducting precision agricultural tasks if equipped with the proper implements (agricultural tools or utensils). Those implements (variable application rates of fertilizers or sprays, mechanical intrarow weed control, and seed planters) are also being automated with the same types of sensors and actuators used in autonomous vehicles (GPS, machine vision, range finders, etc.) [6–11]. Thus, when integrating a given vehicle and a particular implement, many sensors and/ or actuators are duplicated and, worst of all, a central, external computer must be used to coordinate the arrangement: vehicle and implement. Minimizing the hardware of the vehicle-implement system is essential for commercializing reliable, efficient, and cost-competitive agricultural machinery [12]. Thus, devising a simple controller for both vehicle and implement would facilitate reliability, efficiency, and competitiveness.

Many research groups are developing specialized autonomous applications for agriculture that will be operative in the coming years [13–15], but many others are also aiming to operate a group of vehicles under unified control. This is the emergent concept of fleets of robots, which represents a step forward in agriculture. The associated theoretical foundations fleets of robots have been investigated recently [16, 17], but the first applications for agriculture are currently under development [18, 19]. For this purpose, the concept of reducing redundant devices coordinating different, heterogeneous systems by using a central, external computer is prominent.

Fleets of robots can provide many advantages [20–23]: using a group of robots cooperating with each other to achieve a well-defined objective is an emerging and necessary concept to achieve the application of autonomous systems in daily agricultural tasks. The implementation of complex and expensive systems will be attractive for high-value crops for which smart machines can replace extensive and expensive repetitive labor. However, for a robotic agricultural application, considerable information must be processed, and a wide number of actuation signals must be controlled, which may present a number of technical drawbacks. Thus, an important limitation is that the number of total devices (e.g., sensors, actuators, and computers/controllers) increases according to the number of fleet units, and thus the mean time between failures decreases drastically because a failure in one robot component causes the entire fleet to be out of order. This decrease in the time between failures significantly influences fleet reliability, which is of paramount importance for the application of automated systems to real tasks and, in particular, to agriculture.

To achieve a flexible, reliable, and maintainable fleet of autonomous mobile robots for agricultural tasks, the system architecture (involving sensors, actuators, and the computers performing the algorithms) for both the vehicle navigation system and the operation of the implement must be robust, simple, and modular. One of the most important tasks in a control configuration design is the selection of the number and type of sensors, actuators, and computers. These components constitute the basis for the design of the architecture and are very difficult to decrease in number because the processes of perceiving and actuating cannot be avoided; however, these sensors and actuators are typically handled by independent controllers, specifically, commercial off-the-shelf (COTS) sensors such as LIDARs and vision systems. However, computers are sufficiently flexible to share resources and improve the robustness of the system.

In fully autonomous agricultural systems, several actions must be executed simultaneously to ensure effective application as well as safety (including the system, the crop field, and external elements, e.g., human supervisors). Absolute or relative localization in the field, obstacle and interesting element detection, communication with external users or with other autonomous units, autonomous navigation or remote operation, and site-specific applications are some of these specific actions that, all together, compose a fully autonomous agricultural system. This system can be divided into two main subsystems (see Figure 1): the autonomous vehicle and the autonomous implement. The autonomous vehicle, such as a modified commercial tractor, specialized platform, or small vehicle, guides the agricultural system in a crop field for the purpose of executing a crop operation (e.g., harvesting, hoeing, and weed control), which will be accomplished by the autonomous implement. Given the complexity of the assignment, a large number of specialized sensors and actuators are required to fulfill the given task in the given environment.

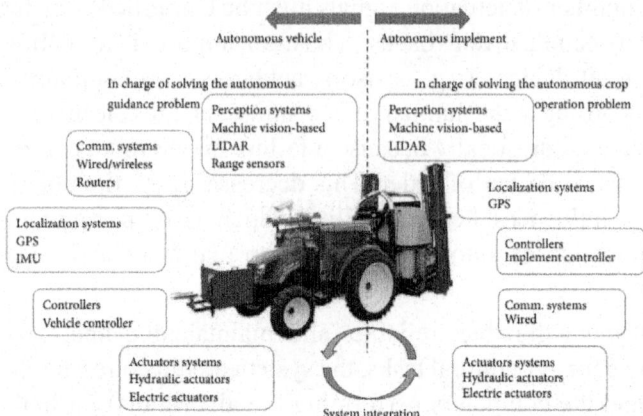

Figure 1: Main systems comprising a current autonomous agricultural application and some examples of sensor and actuation systems normally found in this type of application.

For each individual system presented in Figure 1, intensive research activities have been documented in the literature that intend to solve both the autonomous guidance problem and the autonomous crop operation problem individually. Table 1 presents selected examples of efforts to solve the autonomous guidance problem, and Table 2 presents some works focused on solving the autonomous crop operation problem, indicating the application for which they were developed and the main sensor system used.

Table 1: Examples of autonomous vehicles for agricultural applications developed around the world

Author/Centre	Blackmore et al., 2004. Dept. of Agricultural Sciences, Frederiksberg, Denmark [24]
Application	Automatic steered tractor capable of following a predefined route plan
Sensorial System	RTK GPS: Localization
Results	The automatic steered tractor can follow a predetermined route within a few centimeters
Author/Centre	Cho and Lee, 2000. Department of Agricultural Engineering, Seoul National University, Korea [55]
Application	Autonomous operation of a speedsprayer in an orchard (a speedsprayer is defined as a power sprayer used to apply a highly concentrated pesticide in highly dispersed form by delivering it into a strong air blast generated by fans or blowers—Merriam-Webster Dictionary)
Sensorial System	DGPS for Localization; ultrasonic sensor for obstacle detection
Results	Speedsprayer autonomous operation: within 50-cm deviation. The speedsprayer could avoid trees or obstacles in emergency situations
Author/Centre	Hague et al., 2000. Silsoe Research Institute, Wrest Park, UK [56]
Application	Ground-based sensing methods for vehicle-position fixing
Sensorial System	Sensor package: machine vision, odometers, accelerometers, and a compass
Results	Reducing the low noise level of the odometric data and eliminating drift using sensor fusion
Author/Centre	Subramanian et al., 2006. Department of Agricultural and Biological Engineering, University of Florida, USA [57]
Application	Autonomous guidance system for use in a citrus grove
Sensorial System	Machine vision and laser radar (LADAR)
Results	Machine vision guidance: average error of 2.8 cm. LADAR guidance: average error of 2.5 cm (Tested in a curved path at a speed of 3.1 m/s)
Author/Centre	Xue et al., 2012. Department of Agricultural and Biological Engineering, University of Illinois, USA [58]
Application	Variable field-of-view machine vision method for agricultural robot navigation between rows in cornfields
Sensorial System	Machine vision with pitch and yaw motion control
Results	Maximum guidance error of 15.8 mm and stable navigational behavior

Table 2: Examples of autonomous implements for agricultural applications developed around the world

Author/Centre	Blasco et al., 2002. Instituto Valenciano de Investigaciones Agrarias (IVIA), Spain [59]
Application	Non-chemical weed controller for vegetable crops
Sensorial System	Two machine vision systems: one in front of the robot for weed detection; the other for correcting inertial perturbations
Results	The system was able to eliminate 100% of small weeds. The system properly located 84% of weeds and 99% of lettuces
Author/Centre	Lee et al., 1999. Biological and Agricultural Engineering, University of California, USA [60]
Application	Real-time intelligent robotic weed control system for selective herbicide application to in-row weeds
Sensorial System	Two machine vision systems: one in front of the robot for guidance; the other for weed detection
Results	24.2% of the tomatoes were incorrectly identified and sprayed, and 52.4% of the weeds were not sprayed
Author/Centre	Leemans and Destain, 2007. Gembloux Agricultural University, Belgium [61]
Application	Positioning seed drills relative to the previous lines while sowing
Sensorial System	Machine vision for guidance
Results	The standard deviation of the error was 23 mm, with a range of less than 100 mm
Author/Centre	Pérez-Ruiz et al., 2012. University of California, Davis, Department of Biological and Agricultural Engineering, USA [8]
Application	Automatic mechanical intra-row weed control for transplanted row crops
Sensorial System	RTK-GPS for controlling the path of a pair of intra-row weed knives
Results	A mean error of 0.8 cm in centering the actual uncultivated close-to-crop zone about the tomato main stems, with standard deviations of 1.75 and 3.28 cm at speeds of 0.8 and 1.6 km/h, respectively

A few attempts to establish a fully autonomous agricultural system by integrating an autonomous vehicle and an autonomous implement can be found in the scientific literature. One of the most important examples is the work conducted in Denmark by Nørremark et al. [14]. These authors developed a self-propelled and unmanned hoeing system for intrarow weed control comprising an autonomous tractor [24] and a cycloid hoe [25] linked via a hydraulic side-shifting frame attached to the rear three-point hitch of the tractor. In this system, the autonomous tractor follows a predefined route parallel to the crop rows and turns at the end of the rows, the side-shift frame adjusts its lateral position depending on predefined waypoints, and the cycloid hoe controls the tines to avoid contact with crop plants. Both the vehicle and the implement are controlled independently according to a predefined mission. However, some sensorial systems are replicated; for example, there is one GPS for the vehicle guidance and another for the side-shifting and cycloid hoe control systems.

Some authors agree that, in general terms, the framework of an agricultural autonomous guidance system mainly consists of four subtasks: sensor acquisition, modeling, planning, and execution [3, 26]. Based on this generalization, Figure 2(a) presents a simplified framework for agricultural guidance in which the outputs of each subtask are highlighted. Following this framework, and based on a review of the research activities in autonomous crop operations over the last fifteen years, we can construct an analogy and present a general framework of agricultural autonomous implements (see Figure 2(b)).

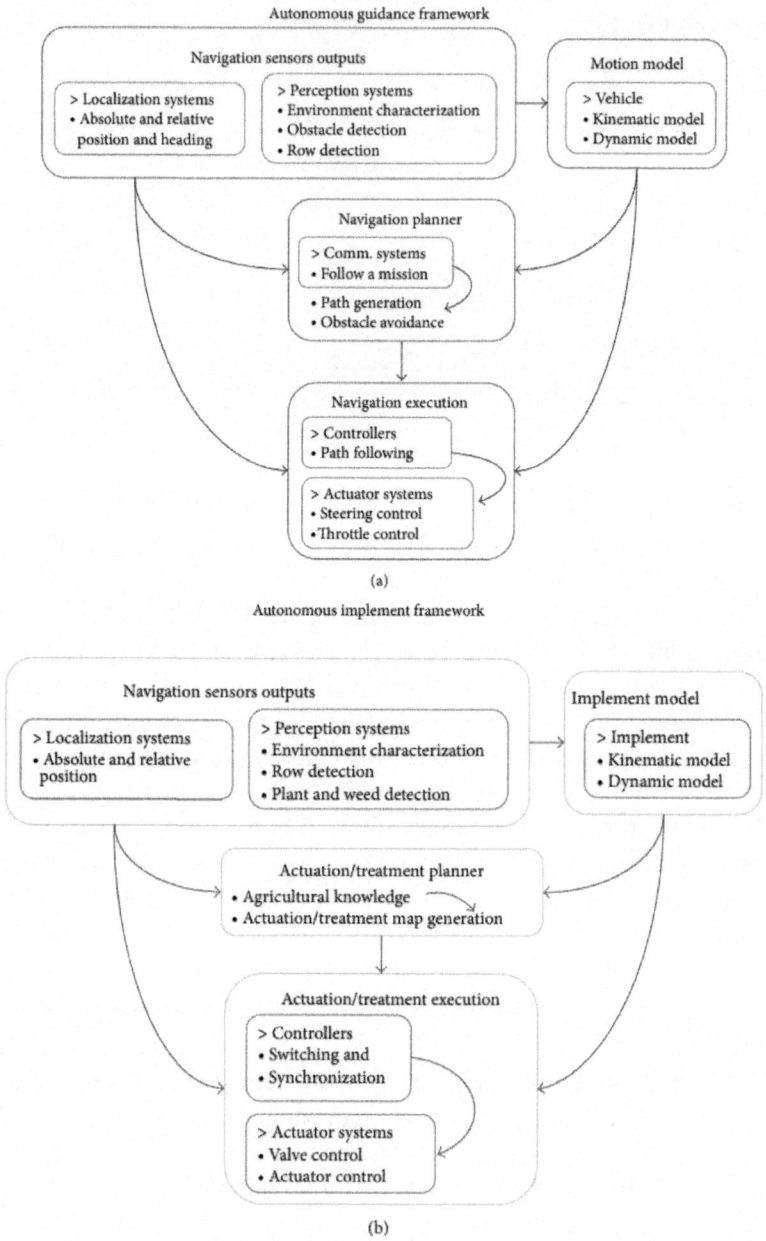

Figure 2: General frameworks of a fully autonomous crop operation. (a) Basic elements of agricultural vehicle guidance systems. (b) Basic elements of autonomous implements.

For each framework, we can identify some similarities in (a) the usage of sensors and actuators, (b) the flow data general scheme, and (c) the specific subtask that uses the sensors and actuators. For example, the use of machine vision in both frameworks is commonly applied for row detection to localize and adjust the relative position of the vehicle/implement depending on the environment; the use of the GPS in both frameworks is commonly applied for absolute localization to follow a predefined route or for the application of a specific treatment in a specific location.

PROBLEM APPROACH

To obtain a fully autonomous agricultural system, the aforementioned two general frameworks must be merged in an architecture (hardware and software) that shares the sensorial system and the planning methods for both the autonomous guidance and the autonomous treatment application. This task must be performed with the objective of reducing the amount of hardware while maintaining the required performance. This architecture must be capable of integrating different sensor and actuation systems developed by diverse research groups as well as different types of commercial equipment. Furthermore, it must be flexible and integrate several standard communication protocols that are common in high-tech agricultural applications [27]. A modular architecture to provide convenient settings of the interfaces between the sensors and devices and proper organization of the perception, processing, and actuation of these types of systems are required due to the large variety of available technologies.

Thus, as a first step, this paper focuses on the design of a proper structure for mobile autonomous vehicles collaborating as a fleet of robots in agricultural tasks. Hardware reliability, truly plug-and-play features, and programmability are essential for efficient agricultural vehicles and, consequently, for competent fleets of robots, but modularity, expandability, ergonomics, maintenance, and cost are also of paramount importance to increase the number of prospective real applications in agriculture.

The aforementioned basic features are considered in the proposed configuration; however, other features are also discussed in the following sections with the primary aim of providing manufacturers of agricultural machinery with solutions for automating new developments, particularly in precision agriculture, an emerging area demanding robust and efficient solutions.

The work presented in this paper has been conducted within the RHEA project, a FP7 program project granted by the European Commission. RHEA is focused on the development of a new generation of vehicles for effective chemical and mechanical management of a large variety of crops to minimize the use of agricultural inputs to decrease environmental pollution while improving crop quality and safety and reducing production costs [21]. To accomplish this aim, RHEA is conducting research in (a) advanced perception systems to detect and identify crop status, including crop row detection, and (b) innovative actuation systems to apply fertilizers and herbicides precisely as well as to remove or eliminate weeds directly. Additional research is focused on the development of (c) a fleet of small, safe, reconfigurable, heterogeneous, and complementary mobile units to guarantee the application of the procedures to the entire operation field. This scientific activity must be complemented with technical developments in (d) novel communication and location systems for robot fleets, (e) enhanced simulation systems and collaborative graphic user interfaces, and (f) pioneering fuel cells to build clean and efficient energy sources (see Figure 3).

Figure 3: The RHEA fleet (ground mobile units and implements).

To accomplish these overall objectives, we have developed the structure presented in this paper, which is organized as follows. First, the architecture of an autonomous system is introduced in Section 3; in Section 4, we collect the requirements for agricultural fleets of robots; different topologies for fleets of robots are discussed in Section 5; finally, Section 6 presents some results, followed by conclusions in Section 7.

STRUCTURING A FULLY AUTONOMOUS AGRICULTURAL SYSTEM

The first idea that comes to mind to structure a fully autonomous agricultural application is to take two operational subsystems such as those presented in the previous section (one autonomous vehicle and one autonomous implement) and put them to work together. This demands a communication mechanism between the autonomous vehicle and the autonomous implement in the form of a Main Controller, that is responsible for merging the desired behavior of each individual subsystem into a single behavior that treats the fully autonomous agricultural system like a robot unit. Thus, the whole system can be broken down into three main modules: vehicle, implement, and controller.

The Vehicle

The vehicle is the module in charge of ensuring the motion behavior of the implement (absolute position and orientation) and must adapt both the type of crop and the type of operation on the crop. Normally, the vehicle carries or tows the implement and therefore provides the necessary energy to the implement as well. Thus, the vehicle must include mechanical adaptors (three-point hitch) to fulfill agricultural standards, electrical generators, and hydraulic pumps. These specific subsystems are provided by commercial agricultural vehicles, and thus adapting a commercial agricultural tractor to configure an autonomous vehicle is easier and more efficient than developing an agricultural robot from scratch. This also allows the developers to advance system integration and testing while avoiding other time-consuming activities such as chassis design, manual assembly, testing, and vehicle homologation, for instance. These modifications drastically increase vehicle reliability by using long-term tested items (engine, braking, steering and transmission systems, housing, etc.) while decreasing time until availability. The safety, robustness, and efficiency of the system must also be considered when structuring the entire autonomous system.

The final selected vehicle for the RHEA project was a CNH Boomer-3050 (51 hp—37.3 KW, 1200 kg), whose restructured and empty cabin was used to contain the computing equipment for the perception, communication, location, safety, and actuation systems. In addition, some systems require the placement of specific elements outside the cabin: vision camera, laser, antennas (GPS and communications), and emergency bottoms. This overall equipment can be classified into the following subsystems:(i)a Weed Detection System to detect weed patches that relies on machine vision;(ii)a crop row detection system to help steer the vehicle based on machine vision;(iii)a laser range finder to detect obstacles in front of the mobile units;(iv)communication equipment linking the operator station, the mobile units, and the user portable devices;(v)a two-antenna global positioning system to locate/orientate the vehicle in the mission field;(vi)an inertial measurement unit (IMU) to complement the GPS data and enable improved vehicle positioning;(vii)a vehicle controller in charge of computing the steering control law, throttle, and braking for path tracking purposes. Steering, throttle, clutching, and braking are the mechanisms normally provided by modern vehicles via a CAN bus;(viii)a central controller as a decision-making system responsible for gathering information from all perception systems and computing the actions to be performed by the actuation components;(ix)an additional energy power supply based on a fuel cell, which is monitored by the central controller.

Figures 4(a) and 4(b) illustrate the original and modified Boomer T3050, respectively. The latter image shows the reduced cabin, the fuel cell, the solar panel placed on top of the robot, the antenna bar, and the equipment distribution inside the cabin. These two last elements are magnified in Figures 4(c) and 4(d).

Figure 4: (a) Initial commercial tractor, (b) final RHEA mobile unit, (c) external equipment onboard the mobile units, and (d) internal equipment distribution inside the mobile unit's cabin.

Implements

The implement is a device designed to perform an action on the crop, such as herbicide and pesticide booms and mechanical and thermal weed removers. The nozzles and burners found on implements are normally operated independently to focus the actuations according to precision agricultural principles. Some of those elements have positioning devices to improve treatments. PLCs and computers are used to control those independent elements and coordinate actions with the vehicle.

The RHEA project has developed three different implements so far.

A boom sprayer [28] for herbicide application in cereals (see Figure 5(a)) consists of a 5.5 m boom containing 12 nozzles separated by 0.5 m and exhibiting independent actuation. The implement is carried by the vehicle and contains two herbicide tanks (200 L and 50 L, resp.), the contents of which can be mixed to apply different treatments. The flow of herbicide through the nozzles as well as the boom folding/unfolding device is controlled by the vehicle's Main Controller.

(a)   (b)   (c)

Figure 5: Implements controlled by the RHEA system: (a) boom sprayer, (b) flame hoe, and (c) canopy sprayer.

A mechanical-thermal machine [29] is for weed control (see Figure 5(b)) in flame-resistant crops such as maize, onion, and garlic. This system consists of four couples of burners attached to a main frame that tackles four consecutive crop rows. The implement is towed by the vehicle, which is also responsible for controlling the relative lateral position of the implement with respect to the vehicle's position. The flame intensity of each burner is a function of the amount of weeds detected by the Weed Detection System based on machine vision. That amount is expressed as the percentage of the area covered by weeds in every area unit—typically 0.25 m × 0.25 m. The vehicle's controller is also in charge of the folding/unfolding device.

An airblast sprayer [30] for pesticide application in olive trees (see Figure 5(c)) consists of two vertical booms with four nozzles each. The lower and upper nozzles are oriented by stepper motors based on the information provided by a set of ultrasound sensors, one per nozzle, with the aim of maximizing the amount of pesticide applied to the canopies. The vehicle passively tows the implement, which contains all of the sensorial systems required for the aforementioned application.

The aim of this section is simply to illustrate the large number of different types of sensors and actuators used in these implements. Thus, the detailed aspects of these designs are considered outside of the scope of this paper.

Main Controller

The Main Controller is in charge of steering the vehicle accurately, coordinating the actions of the vehicle, and maintaining communication with the operator. In addition, the Main Controller integrates a large number of subsystems, such as those mentioned in Section 3.1. Integrating different systems based on diverse communication technologies, operating systems, and programming languages leads to questions about the organization of the hardware and software architecture, which can be centralized or distributed, open-source or commercial development software, among others. These options have advantages and disadvantages that can be found in any technical material on the topic. A distributed architecture is based on several computers running different applications on similar or dissimilar operating systems. The computers are connected by a communication network or point-to-point communication links exhibiting very well-known features as well as a few shortcomings derived from its maintenance cost in terms of updates and security, the number of different operating systems and programming languages to be handled, time-consuming management needs, and network delays in communicating data, which may impair the real-time features of the system.

Apart from considering the advantages and limitations of both configurations, the optimum choice depends on the specific application, that is, the number of sensors; the number and type of peripherals; the number of different computers, including operating systems and languages used; the required computing power; and the real-time requirement, among other factors. This task is relatively easy to perform in a closed requirement system, that is, a system in which we know the exact number of subsystems and their features. However, in agriculture, the number of different system configurations, the available commercial devices, and custom-made equipment make the selection of the optimum configuration a difficult task, particularly due to the different operating systems and programming languages.

The best solution, as in other engineering fields, could be to use a hybrid architecture featuring centralized and distributed characteristics capable of integrating new systems when possible and permitting the connectivity of the complex system by using distribution features, such as Ethernet networks and a CAN bus, among others.

In the last ten years, developers of robotic systems, particularly universities and research centers, have been attempting to consolidate and package robotic frameworks as open-source software available to the entire robotic community. Examples of these frameworks are MOOS [31], PLAYER [32], CARMEN [33], and ROS [34], among others, which are essentially network-based communication architectures that allow diverse nodes or applications to communicate and interact with each other. These applications are packages developed by other research groups or by the users, and they are commonly used in the academic community and research centers and commonly applied in service robots. Currently, the most popular open-source operating system for robots is ROS (Robot Operating System), a software platform comprising a large collection of open-source libraries and tools that was initiated in 2007 for the development of robot software and provides the functionality of an operating system on a heterogeneous computer cluster. This system provides standard operating system services (hardware abstraction, device control low-level message passing between processes, implementation of commonly used functions, and package management). ROS is the reference in many research and academic developments because it is free and powerful, but it is released under the terms of the Berkeley Software Distribution licenses, a family of permissive free software licenses that imposes minimal restrictions on the redistribution of covered software, complicating its application to systems for commercial exploitation [35]. Although packaging a robotic framework facilitates the integration of systems from different providers with very dissimilar features, it also leads to problems related to revealing the expertise (making the software code available to others may cause replication and loss of financial benefit) and loss of income through sales (revenue must be gained through support agreements and OEM customization). Recently, some have suggested that ROS should be locked down, protected, and commercialized [36] to monetize industrial and service robots [37].

IDENTIFYING ARCHITECTURE REQUIREMENTS FOR AGRICULTURAL FLEETS

Fully Autonomous Agricultural System Requirements

As presented in Sections 2 and 3, an important aspect of structuring an architecture for a fully autonomous agricultural system (vehicle and implement) is the reduction of the amount of sensors and actuators of the entire system, which constitute the basis for the design of the hardware architecture. However, decreasing the amount of devices for sensing and acting is a difficult task because these components are needed for the correct operation of the system.

Analyzing the two general frameworks presented in Figure 2 reveals that some tasks for guidance and actuation require similar sensorial systems and similar information processing, particularly the tasks of localization, perception, and planning. Furthermore, in a fully autonomous system, instead of having two processes for each of the aforementioned tasks, which would replicate hardware and software elements, these similar tasks can be merged to reduce the amount of specialized hardware.

When merging the tasks of each individual subsystem, the problem of resource assignment and synchronization arises. In addition, the vehicle and implement move according to different reference frames, but a general behavior of the fully autonomous agricultural system must be determined as a part of the general mission of the entire fleet of robots.

Another key element of the hardware architecture is the ability to allow diverse vehicle and implement configurations to enable a fleet of heterogeneous robots to execute diverse crop operations at the same time. To achieve this capability, the hardware architecture must be modular to allow diverse sensory and actuating elements to be rapidly and easily replaced, installed, and configured, thus modifying a small part of the fully autonomous system to enable diverse crop operations. The link between sensors and actuators relies on the computer system.

Given this preliminary discussion, agricultural fleets of robots should rely on the following elements.(i)A hybrid computing system consisting of a central, powerful, truly real-time, multitasking computer with fast network communication features to connect different peripherals.(ii)The central computer should have a large family of real plug-and-play hardware modules including both reliable wired and wireless communication modules. (iii)The central computer should provide capabilities to facilitate running software developed for different platforms and in different languages.(iv) Simple and powerful connections to external libraries and third-party tools must be included.(v)The development tools must allow diverse programming languages for different applications and domain experts in different disciplines (e.g., agronomists and roboticists) and must permit multidisciplinary use, for example, a graphic programming system.(vi)The central computer should allow a wide variety of data acquisition and embedded control devices, which tightens software-hardware integration.(vii)The central computer should be ruggedized to operate in harsh conditions and allow intrinsic parallelism: multicore-ready design and support for different hardware acceleration technologies: DSPs (Digital Signal Processing), FPGAs (Field-Programmable Gate Array), and GPUs (Graphic Processing Units) as coprocessors.(viii)The central computer must have the capability to execute and solve complex algorithms in real-time using real-world external signals (A/D).(ix)The entire architecture must be able to transition easily from academics to industry, ensuring protection of property rights.

Many of these features are fulfilled by the new family CompactRIO-9082 (cRIO-9082: 1.33 GHz dual-core Intel Core i7 processor, 32 GB nonvolatile storage, 2 GB DDR3 800 MHz RAM), high-performance integrated systems commercialized by National Instruments Corporation, whose equipment has already been used in some unmanned road vehicles [38, 39] and autonomous agricultural vehicles [13]. The selected system offers a powerful stand-alone and networked execution for deterministic, real-time applications. This hardware platform contains a reconfigurable Field-Programmable Gate Array (FPGA) for custom timing, triggering, and processing and a wide array modular I/O for any application requirement. This system is designed for extreme ruggedness, reliability, and I/O flexibility, which is appropriate for the integration of different sensorial and actuation systems in precision agriculture autonomous applications.

Furthermore, LabVIEW (Laboratory Virtual Instrumentation Engineering Workbench) is a graphical programming environment used to measure, test, and control different systems by using intuitive graphical icons and wires resembling a flowchart. This environment facilitates integration with thousands of hardware devices, provides hundreds of built-in libraries for advanced analysis and data visualization, and can help prototype applications with FPGA technology.

These hardware/software features ensure (a) performance: equipment reliability and robustness in harsh environments; (b) compatibility: a large list of modules is available for peripherals, including serial and parallel standard communications; (c) modularity/expandability: a computer-based system includes a set of configurable modules that allow the system to grow according to the application needs; (d) developer community: the increasing number of LabVIEW users sharing their experiences and developments in the form of packages or functions freely and openly through blogs and forums; and (e) cost: while the investment in NI hardware and software is initially high, profitability must consider the reduction of the development in person-months as well as the reduction of the hardware when manufacturing medium-to-large prototype series by using products such as the Single-Board RIO.

Based on the previous analysis of both hardware and software features, we have proposed the aforementioned system as the Main Controller of the RHEA robot fleet [40]. Key features in the selection of the present controller were the capabilities of configuring a minimum hardware and assuring a short period for software development. These features allow the developers to focus on the implementation of new algorithms and on the integration of sensors and actuators.

Fleet Management Topology Requirements

Once the architecture requirements for the implementation of a fully autonomous agricultural system are defined, it is necessary to define the requirements for the fleet of robot, which comprises several robot units as described in the previous sections, oriented to agricultural applications. Basically, a fleet is a set of independent mobile units that must be coordinated somehow and interfaced with (a) the environment or workspace; (b) with each other; and (c) with an operator, at given instants. In robotic agriculture, the workspace normally is well known: the dimensions of the field or set of fields are well demarcated; the field is planted or needs to be planted with a specific crop; the boundaries and fixed obstacles are well known; and the areas where the vehicles can travel are well determined. In addition, coordinated motion in this workspace involves relatively small teams of both similar (e.g., tractors with sprayers) and/or heterogeneous vehicles, (e.g., harvester and truck), depending on the application and the final goal. Each of these situations leads to diverse solutions of the coordinated motion problem. In some applications, where two or more vehicles must constantly cooperate (e.g., autonomous harvesting), motion coordination between nearby vehicles is more critical to ensure path accuracy, dead distance, time, fuel, or other efficiency criteria than in other applications in which each vehicle performs a defined and repetitive task without cooperation (e.g., transplanting and weed control). Some attempts have been made to solve the coordinated motion problem in the form of conceptual farming system architectures [41–45], including some that have put cooperation between robots in agricultural tasks into practice [46, 47].

However, given the workspace characteristics and the well-defined general objective of the fleet, many authors agree that a central planner running on an external computer that knows each of the elements involved in the agricultural application and is capable of readjusting the parameters and assignments is necessary for optimal development of the general agricultural task. However, depending on the type of agricultural application for which the fleet is configured, each autonomous unit could have a greater or lesser ability to replan its own subtasks. Conceptual examples can be found in [48–50].

Even if the workspace is well defined, safety is an important factor that affects the fleet composition. The vehicles should be in frequent communication with the external computer to provide data about current status and operation, and a human operator must be in constant supervision. The operator must be present at some instants: mission configuration, mission start, and mission stop or suspension, among others. Thus, an operator interface is essential.

Based on these requirements, the topology of the fleet of robots defined for the RHEA project was a central-external computer located in a base station (BS) for planning, supervising, and allowing the user to access a full interface, in addition to a user portable device (UPD) that allows the user to approach the units to maintain control and supervision of the fleet (see Figure 6). In this topology, a master external computer connected to the fleet units through a wireless communication system runs a mission manager (mission planner and mission supervisor) that sends commands to (and receives data from) the fleet mobile units.

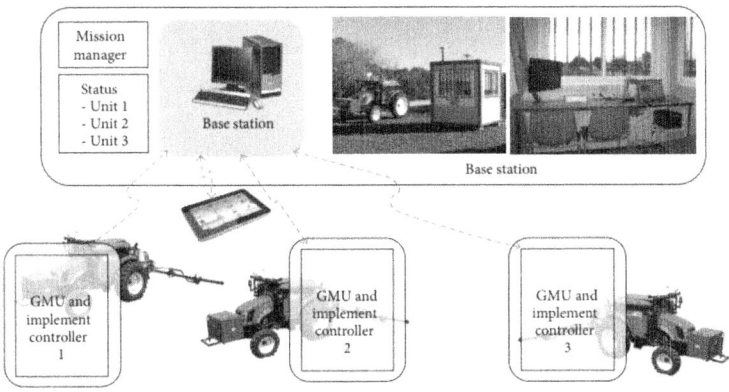

Figure 6: General schema of the fleet of robot topology for the RHEA project.

IMPLEMENTATION OF THE PROPOSED MAIN CONTROLLER: THE EVOLUTION OF THE RHEA COMPUTING SYSTEM

The computing system onboard the mobile units must communicate with a large number of subsystems, such as those specified in Section 3.1, which are based on computers running different operating systems (e.g., Windows, Linux, and QNX) and software modules developed in different languages (C++, NET, Python, etc.). The first solution was to connect all subsystems through an Ethernet network and use a computer as a central controller [27].

This initial solution is depicted in Figure 7. The Main Controller is connected to the peripherals through either a serial line or an Ethernet network (802.3 Local Area Network) via an Ethernet switch, which requires a Network Manager running on a computer connected to the Ethernet switch, normally the Main Controller.

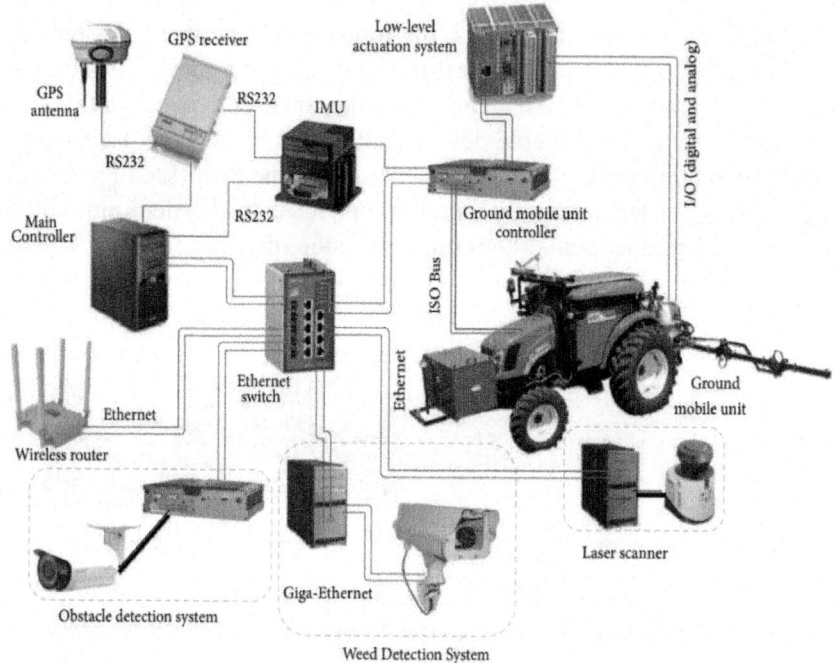

Figure 7: General scheme of the hardware architecture for the autonomous mobile robot in the RHEA project.

The first step toward centralization consisted of integrating the Weed Detection System (WDS) into the Main Controller. The vision camera is GigE Vision standard compliant (global camera interface standard developed using the Gigabit Ethernet communication protocol framework for transmitting high-speed video and related control data), and the Main Controller has two Gigabit Ethernet ports. This allows for a direct interface between the camera and the Main Controller using the functionalities provided by LabVIEW for configuration and acquisition, avoiding the development of new drivers and eliminating the vision computer. This solution is illustrated in Figure 8.

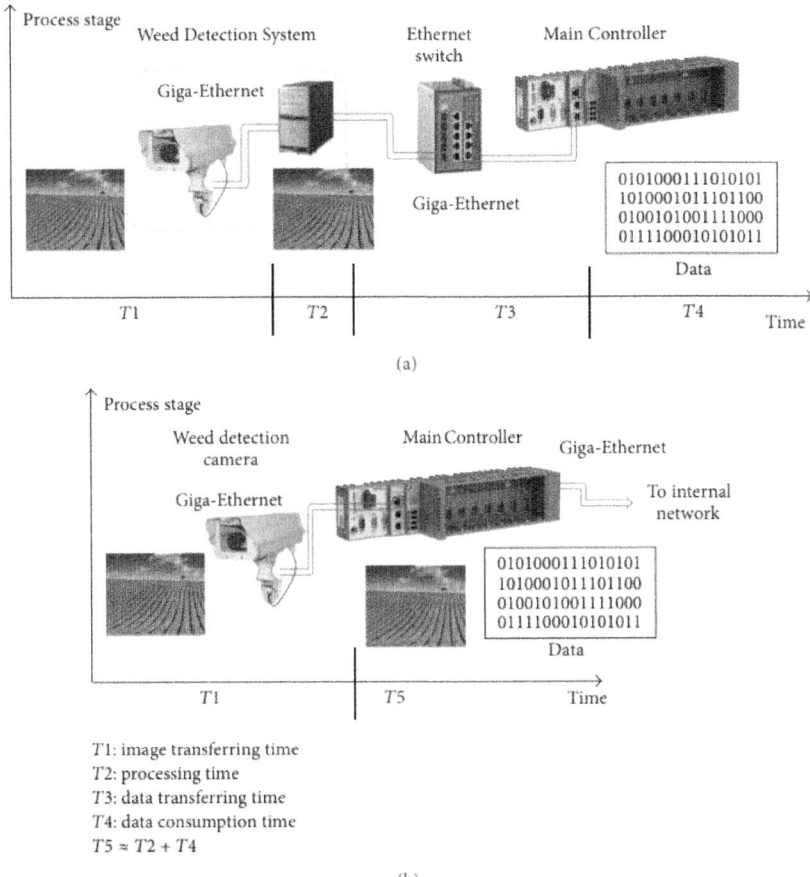

Figure 8: Comparison of the distributed approach (a) and the centralized approach (b) in the Weed Detection System regarding the use of resources, information availability, and communication time.

Two major problems arose at that time. The first was reusing the acquisition software implemented in C++; the second was to assess the Main Controller running the vision algorithms as an additional charge. The first problem was solved by using the LabVIEW connectivity with third-party tools, which allows the programmers to call external scientific libraries in the form of C code, DLLs (Dynamically Linked Library), and code interface nodes (CINs), which include C code compiled into LabVIEW nodes. The specific solution consisted of converting the vision code developed in the C++ language for Windows 7 into a DLL. This DLL can be loaded into the Main Controller and its functions can be called within the LabVIEW environment. One of the important steps in creating a compatible DLL is the detection and substitution of pieces of code that may have problems during the execution, such as system calls. This problem can occur when an external code developed for other operating systems (in this case Windows 7) is called by the LabVIEW Real-Time Operative System (LabVIEW RTOS) and attempts to access some kernel libraries. This may generate conflicts, and therefore it is recommended that this practice should be avoided as much as possible. Once the source code is adapted for execution in the LabVIEW RTOS, it must be packaged in one or different C language functions following the procedure defined in Figure 9.

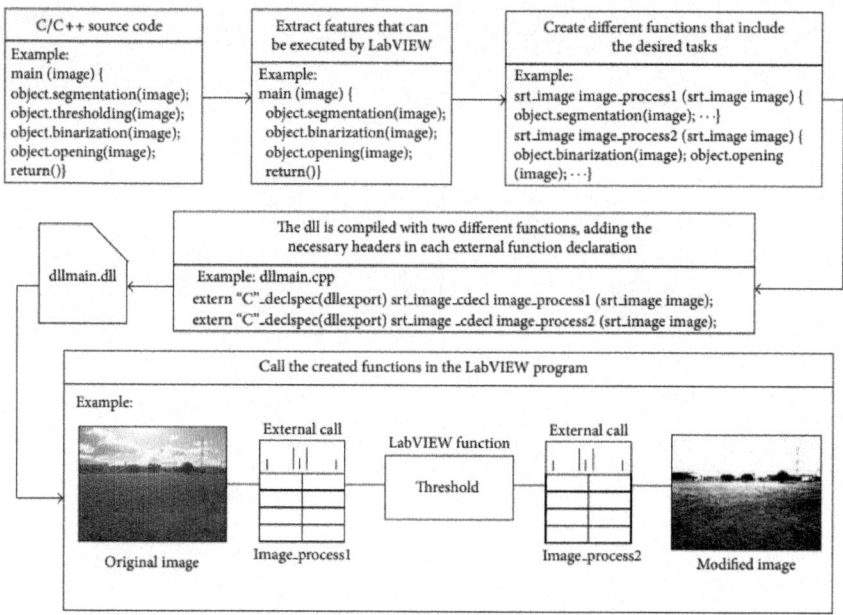

Figure 9: Example of the procedure for calling external code in LabVIEW using DLLs.

There are several advantages of running the real-time features of the algorithms in the same computer (see Figure 8).(i)By eliminating the vision computer (WDS computer) and implementing the execution of the weed detection task in the Main Controller, the deterministic performance is increased, which removes an intermediate non-real-time OS (Windows) and an extra Ethernet network link.(ii)By receiving the information directly from the camera, other processes that are being executed in the Main Controller can rapidly access the images by sharing the same memory space.(iii)By integrating the acquisition and processing algorithms in the Main Controller, the information about the weed distribution is rapidly shared with other processes that require it, which decreases the communication time and increases the real-time response. While this integration generates a problem of interprocess communication by shared memory, it is compensated (in this particular case) by eliminating the communication between different machines. Because there is no need to share images with other computers, the development of drivers for these tasks is eliminated.

A second strategy to improve the centralization of the RHEA system is to unify the two vision systems: the Weed Detection System (WDS) and the obstacle detection System (ODS). Both systems use similar image-capture mechanisms and image-processing algorithms, which can be integrated into the same computer to save hardware resources. The software, which is written in the C++ language for the Linux operating system, is converted to a DLL following the procedure described for the WDS (see Figure 9). The main problem with this configuration is the lack of real parallelism in the execution of the algorithms, which increases the computing time. However, this increase is compensated by the elimination of the delay in the information flow from the ODS to the Main Controller through the Ethernet. Analogous to the integration process of the WDS into the Main Controller, Figure 8 shows that, in the proposed architecture, the time T3 will be eliminated in the data flow of the ODS in comparison with the original schema. In the proposed architecture, the camera acquires an image and sends it via the Ethernet to the Main Controller in charge of both processing the image and making the decision. By contrast, with the original schema, the ODS information must pass through more network elements, increasing transmission time.

Some advantages of integrating the two vision systems are as follows. (i)The application requires only one camera, which reduces the amount of hardware and relevant equipment (the vision fields of both cameras are similar). (ii)The Main Controller allocates the same memory space to share information between the two vision processes (which can take advantage of the processed image to be used by the two different processes).(iii)The two processes can be executed in parallel.

One of the further developments and benefits of this integration is the improvement of the performance in an obstacle detection task in real-time applications by fusing the camera and the laser information. Integrating the sensor acquisition methods and the fusion method in the same computer increases the reliability of deadline compliance, data correlation, and synchronization compared to the original scheme. However, if the data acquisition, both with the camera and the laser, is not performed on the decision-making computer, the nondeterministic features of the Ethernet will reduce the real-time capabilities, expanding the timeframe and producing synchronization problems.

These are two examples of possible system centralization of complex subsystems; however, other subsystems can be centralized in a simpler way by using the plug-and-play features (e.g., Ethernet communication through WLAN modules and switches; laser and inertial measurement units through RS-232 serial modules; industrial communication buses through CAN bus; ISO modules, low-level actuation system through analog and digital I/O modules; etc.). Figure 10 shows the final system scheme, which includes the main external sensorial components.

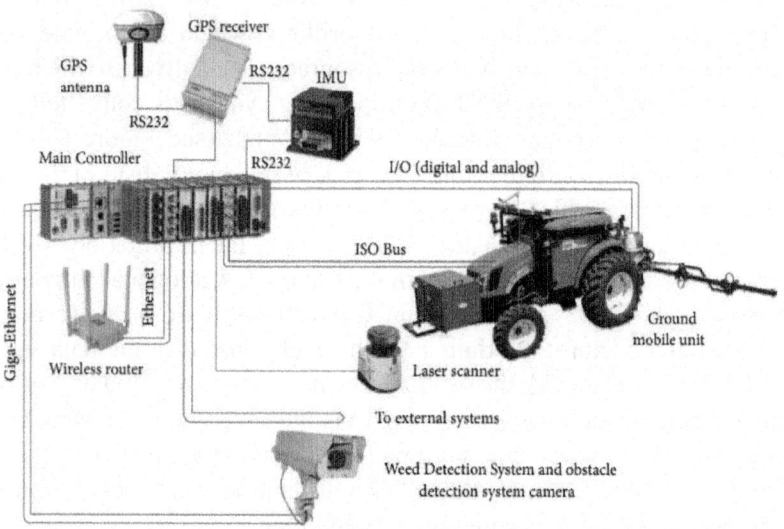

Figure 10: Prospective hardware configuration of the RHEA system.

Using this basic controller and taking advantage of the LabVIEW features, we have defined a simple software architecture to connect all subsystems to the main module in charge of making decisions (high-level decision-making system). Figure 11 illustrates a general schematic diagram in which the following three different software levels are defined.(1)The first level, represented by yellow boxes, consists of driver modules that allow communication with the various sensors, actuators, and other elements of the system (e.g., external user interface).(2)The second level, represented by blue boxes, consists of several modules in charge of interpreting, generating, and/ or merging information from the lowest level to make it more accessible to the decision-making system module or to adjust control parameters for guidance and actuation.(3)In the highest control level, the decision-making algorithm takes the information from the lower-level modules, and based on the desired behavior of the fully autonomous agricultural system, a plan to be executed by the guidance control and the treatment application is formulated.

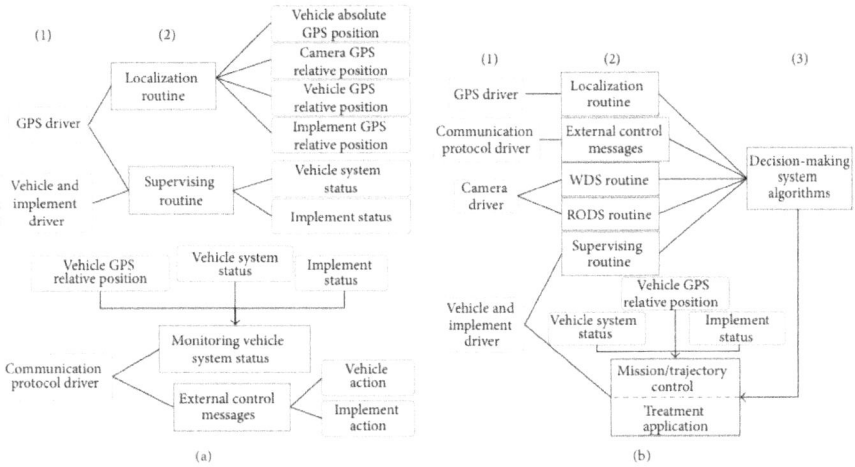

Figure 11: General diagram of the high-level decision-making system indicating three levels of main subsystems, their outputs, and their interactions with other subsystems. (a) Principal outputs (green boxes) of the lower subsystems. (b) Flow between sensory systems and control systems and navigation process execution.

After minimizing the hardware of the individual mobile units, the next step is to minimize the hardware of the whole fleet. The procedure of minimizing the hardware of a fleet of robots relies on the other elements that constitute the fleet of robots: the base station and the operator. As indicated in previous sections, the operator must be present at some instants of the application to configure and supervise the mission. Thus, an operator interface is essential, which can be provided in the form of a base station (computer monitor and keyboard) or a portable device (e.g., tablet, smartphone) that allows the operator to move close to the mobile units. A step forward in the configuration of the fleet of robots would be to structure a fully unmanned fleet with no operator intervention. This prospective case, which is not currently allowed by the legislation of many countries, would dispense with the base station, and the mission manager and the fleet supervisor would be run in the computing system of the mobile units. Two solutions are envisaged as follows.

Master-Slave Configuration

One fleet unit controller acts as a master running the mission manager and the supervisor of the fleet, while the rest acted as slaves receiving commands and returning data. A failure in this master controller also stops the mission of the fleet, but the likelihood of failure decreases because the whole fleet has one less computer and communication system (see Figure 12) with respect to the central-external controller solution. Adaptation to this topology is straightforward: the mission manager algorithms running on the base station computer can be packaged into a DLL and included, with minor software modifications, in a Main Controller, which will act as the fleet master controller. This process is the one explained in Figure 9.

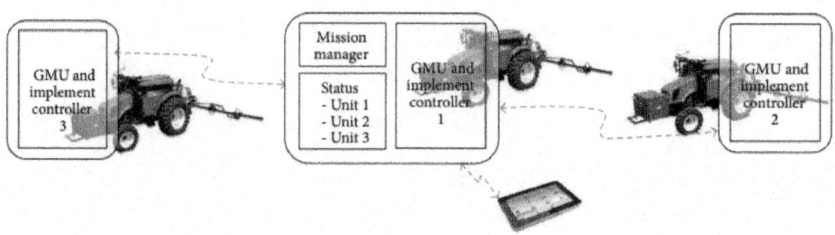

Figure 12: Master-slave configuration.

Besides reducing hardware and structural elements in the fleet of robots, another advantage of this topology is the extension of the working area of the fleet. If the mission supervisor is fixed at a point in the field, the maximum working range of each unit is limited by the range of the communication system. A typical wireless network can have an open field range of up to 150 meters. As indicated in previous sections, the use of a fleet of autonomous robots in agriculture may be feasible in extensive applications that require long hours of continuous working in fields of tens of hectares. Thus, a larger communication range is required. One solution is the use of larger antennas and increasing the power of the transmitter/receiver to maintain an acceptable bandwidth or the use of signal repeaters. However, as there is a master unit in this topology, the mission supervisor has the ability to move around the field, which increases the working area of the entire fleet (maintaining a typical wireless network configuration), as long as the mission is defined so that each unit is within communication range.

Immerse Configuration

The mission manager is copied in all mobile unit controllers so that a failure in one unit means that unit stops, while the rest can reconfigure the mission plan to accomplish the task. Note that the hardware is not increased and that the same mission manager algorithms run on every unit controller (see Figure 13), while the unit statuses are shared among all the robots by broadcasting a few status data in a sampling period basis. When a unit goes out of order, the others receive that information in the status or by a sampling period timeout; in such a case, the remaining active units will compute the mission manager, taking into account that incidence.

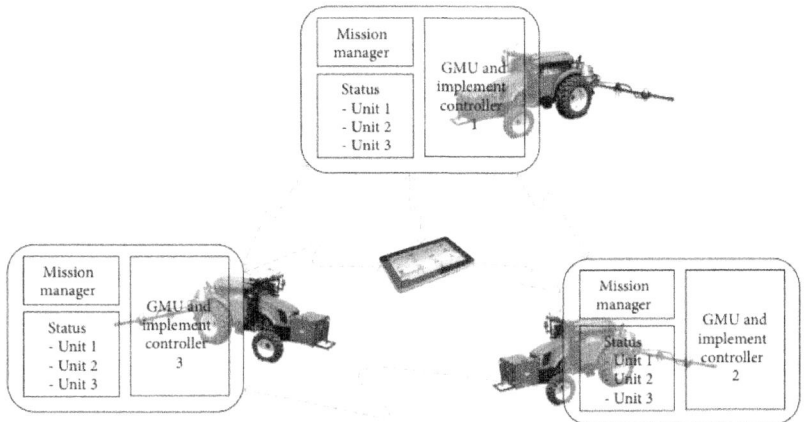

Figure 13: Immerse configuration.

For this solution, there is not a clear gain of hardware reduction in the general architecture, but the immerse controller increases the robustness of the system by using a mirrored mission planner on each mobile unit controller. This immerse controller allows each mobile unit to supervise (mission supervisor) the execution of the plan and monitor the status (position, speed, etc.) of the other mobile units while adapting the missions of individual units to meet the goal of the fleet. This configuration, which is illustrated in Figure 13, is well suited to the hardware architecture presented in Section 5.1 for the ground mobile unit, in which the use of a cRIO system as the Main Controller permits direct communication with other cRIO systems without the development of drivers and communication protocols, thanks to the ability of the LabVIEW utilities to share information between NI systems.

RESULTS

To present the implementation of a working fleet of robots configured with the Main Controller proposed in this paper, a set of assessments was conducted in a real experimental field as part of the RHEA project [40]. Several tests and integrations have been conducted that have positively assessed the system efficiency and ease of new integrations, which are organized as follows: Sections 6.1 and 6.2 present both quantitative and qualitative results associated with both hardware element reduction and software development minimization in a single, fully autonomous agricultural system; Section 6.3 presents the results of an algorithm for collision avoidance, allowing the assessment of the benefit of hardware reduction in a fleet of robots oriented to agriculture.

Integration of the Weed Detection System in the Main Controller

The first assessment trial was focused on evaluating both the image acquisition and processing procedure of the Weed Detection System by using the proposed architecture (see Figure 10) and compared them with those obtained with the original RHEA scheme (see Figure 7). For that trial, we measured the time required for each topology (centralized versus distributed) to acquire an image and generate an output, which was received in the Main Controller (see Table 3). In the first trial (with the original scheme), the computer was exclusively dedicated to image acquisition and image-processing tasks. However, using the proposed architecture, image acquisition, image processing, and four additional tasks defined in Table 3 were executed in parallel, meeting the scheduled time. Considering that each topology generates very similar results, we can conclude that we have maintained the required performance of the system, decreasing the hardware and developing a small number of communication interfaces. Furthermore, the image acquired by the Weed Detection System is available within the Main Controller in half the time using the architecture illustrated in Figure 10 compared to the original scheme (see Figure 7). Because the images are of high resolution, this time is quite significant when performing real-time calculations, and thus the same image can be shared with other processes, such as the obstacle detection system, avoiding redundant hardware (several cameras, for instance).

Table 3: Comparative timing results between the RHEA original schema and the proposed architecture regarding the Weed Detection System.

(a)

Time required	Image Acquisition	Fps acquired	Image Processing	Fps processed	Image Sharing	Other process running
Original structure	75–150 ms	5	150–250 ms	4	150–200 ms	0
Proposed structure	80–160 ms	5	200–250 ms	4	1 ms	4 (See Table 3(b) for process description)

(b)

Other process running	Scheduled periods
Path following supervising routine	100 ms
Steering and throttle control routine	10 ms
Telemetry routine	100 ms
Localization routine	100 ms

Integration of the Ground Mobile Unit Controller in the Main Controller

One more evaluation of the system was performed by removing the ground mobile unit controller (GMUC) in charge of the vehicle guidance and implementing path follower algorithms in the Main Controller. In this case, we evaluated the system capabilities to react to changes in both the trajectory and speed of the vehicle, which were measured as the number of messages sent to control both the vehicle speed and steering. Leaving aside the vehicle mechanical response and the performance of the path-following algorithms, by using the original RHEA scheme, the Main Controller can send messages (new trajectories) at a rate of 6–10 Hz. However, by using the proposed architecture, the Main Controller can send messages (new steering and speed references values) at a rate of 100 Hz. It is not correct to directly compare these two values because the messages correspond to diverse control levels. Therefore, a qualitative analysis must be performed. The original RHEA scheme defines the guidance system as a deliberative architecture in which the trajectory planning is performed by the Main Controller (based on a predefined mission and information of the perception system) and the GMUC executes that plan. The proposed architecture changes this configuration into a hybrid architecture, where, in critical situations (e.g., obstacle avoidance, row guidance, safety procedures), the capabilities of changing the position and orientation of the vehicle are improved. Although the behaviors of these two schemes are well known and have been studied for years [51, 52], they remain a current research topic [53] and are well suited to the requirements defined in Section 4.

The implemented controller relies on a fuzzy logic algorithm developed in [54].

Implementation of a Collision Avoidance Algorithm in the RHEA Fleet of Robots

As a final test to validate the use of the proposed architecture in a fleet of robots oriented to agricultural tasks, the implementation of a method for avoiding collisions between units was evaluated. The fleet configuration was as follows. (a)Regarding each individual, fully autonomous agricultural system, the positioning system (RTK-GPS) was the only sensory element enabled for this test, in addition to the communication system (wireless communication) and the Main Controller (in charge of executing the mission and communicating with both the user and the mission supervisor).(b)The fleet topology used in these tests was the master-slave configuration (see Figure 11), in which the algorithms to configure and transmit the mission to each unit, in addition to the fleet supervisor algorithm, were executed within the Main Controller onboard the GMU1. These algorithms had a built-in user interface that could be accessed remotely by the user via an external computer connected to the network of the fleet. With this interface the user can (a) monitor the status and location of each unit; (b) load a predefined mission to each unit; (c) record the GPS relative position of each unit; and (d) start, stop, or pause the motion of each unit, among other actions.

Each unit must follow a user-predefined path at a constant speed. Each path consists of crossing a real field (35 meter long by 25 meter wide), making a U-shape turn, and returning down the field on a different crop line. The units are allowed to make the turns in an area with a length of approximately 8 meters at the headlines. The GPS positions recorded by each unit and the general mission sent to the fleet of robots are illustrated in Figure 14. Higher concentrations of GPS positions in the figure (the colored circles for each unit) indicate that the unit was moving at low speed or even stopped; by contrast, a lower concentration of GPS positions corresponds to normal development of the submission, following a predefined path at a constant speed.

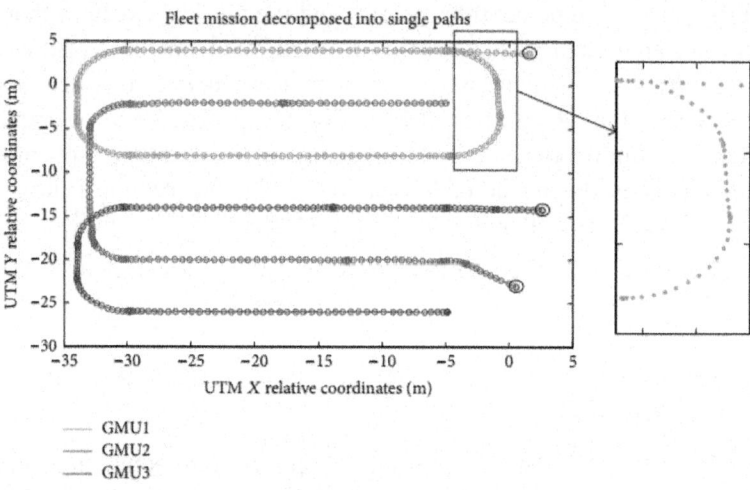

Figure 14: GPS position recorded for each unit representing the mission execution. The black circles represent the origin point of each unit.

Although the fleet mission can be defined as optimal both in time and space (because the characteristics of the field and the fleet are well known), it is possible to identify random external elements that alter the planned development of the mission and generate potential collision situations accordingly. Examples of these situations include the following: (a) detection of moving obstacles (e.g., persons, animals, and other tractors); (b) treatment parameters that affect the speed of operation (e.g., in a weed control treatment in which a decrease in speed is required for a more optimal application); (c) small temporary failures in the system itself (e.g., loss or decrease of the accuracy of the GPS signal, wireless communication loss).

To avoid collisions between units, the fleet supervisor algorithm contains a procedure that receives the GPS positions of each unit and calculates their possible location in subsequent time instants based upon their intended movement (current heading). The collision avoidance algorithm models each tractor as a square element and its intended motion as a conic section in which the vertex of the cone is in the center of each tractor. The opening angle of the conic section depends on whether the tractor is inside the field (smaller angle) or in the headlines (bigger angle), given that inside the field each tractor normally moves along a straight line. The fleet supervisor assigns priorities for each unit to continue its submission or stop until the risk of collision disappears. For this particular case, GMU1 has the highest priority, while GMU3 has the lower priority. The method for detecting potential collisions is occupancy grid mapping (see Figure 15).

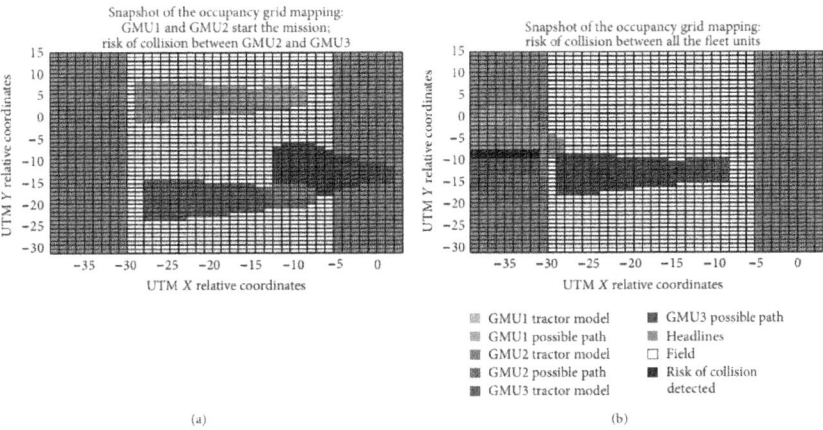

Figure 15: Snapshots of the occupancy grid mapping for collision detection. (a) Results of the collision detection within 10 seconds of the execution time of the mission. (b) Results of the collision detection within 25 seconds of the execution time of the mission.

Figure 16 illustrates the distance traveled by each fleet unit as a function of time. At some times (e.g., in the first 20 seconds of the mission; between the second 25 and 40 seconds), some units remain stopped because the fleet supervisor paused the execution of the submission of these units because there was a potential collision situation. Figure 15(a) shows the result of the collision detection algorithm for an instant of time between the first 20 seconds of the general mission, during which a possible collision between GMU2 and GMU3 is present, and thus GMU3 will take longer to get to the other end of the field. This is the situation presented in Figure 15(b), in which GMU1 and GMU2 are making the turn to return to the field and another possible collision situation is present. In this situation, the fleet supervisor allows GMU1 to continue with its submission while stopping the movement of GMU2 until the collision situation disappears. In addition to the tests conducted for collision avoidance using the master-slave configuration, tests were performed with the original RHEA project configuration (see Figure 6), and as expected, the same results were obtained. These results confirm the potential of the proposed control architecture for an autonomous fleet of robots to allow hardware and software development reduction while maintaining the desired performance.

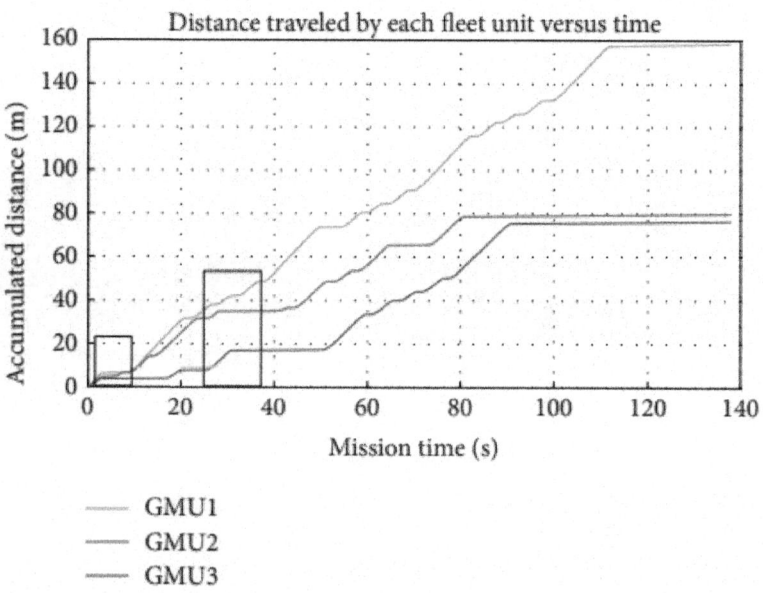

Figure 16: Accumulated distance traveled by each fleet unit as a function of the mission time.

A video of the RHEA fleet is available at http://www.car.upm-csic.es/fsr/gds/RHEA_fleet.mp4eg; the video includes images of the user interface described in Section 6.3 as well as a real-time result of the collision avoidance algorithm.

CONCLUSIONS

Robotics and new technologies have begun to improve common practices in agriculture, such as increasing yield performance and decreasing the use of chemicals that may affect the environment. Furthermore, new robotics systems for application in agriculture are under development to permit the integration of different technologies while enabling modularity, flexibility, and adaptability.

This paper presents a structure for agricultural vehicles to work both independently and in fleets, that is, simple, robust, and reliable. The general scheme exhibits advantageous features to quickly implement new vehicle controllers and develop/integrate advanced agricultural implements. Three examples are reported herein: a boom sprayer, a mechanical-thermal machine, and an airblast sprayer.

The proposed architecture for the centralization of the Main Controller and the principal sensory systems provides some advantages for a future sensor fusion. Integrating critical sensors in autonomous agricultural applications, such as high-definition cameras and lasers systems, allows the information to be merged to improve the performance of the sensory system in terms of greater accuracy, greater robustness, and increased complementary data and to reduce the amount of hardware, which increases the communication speed and the information shared by different modules.

In addition, in an autonomous agricultural application, when the environment exhibits changing light, soil, and crop characteristics, among other characteristics, the sensory system is required to perform more complex tasks, which consequently leads to the problem of overcharging the Main Controller due to both the execution of multiple tasks in the same controller and the high consumption of resources for sensory fusion tasks. Nevertheless, in the proposed solution, this overuse is compensated by the Main Controller characteristics and its ability to execute diverse processes in parallel and in real-time as well as the possibility of implementing very specific and time-critical operations in the FPGA device.

This study attempts to increase the robustness of autonomous agriculture robots and fleets of robots by reducing the equipment hardware onboard the mobile units and facilitating the integration of different sensors devices and software modules developed by professionals in different fields and skills. Moreover, minimizing user involvement in monitoring and safety functions and enabling the same elements of the fleet to manage certain critical situations can also permit the reduction of the amount of hardware and structural elements in the fleet, which might increase the working area of the entire fleet.

The system is operational, and both individual and fleet robot features have been tested. The previous section illustrates two examples of subsystem integration into the Main Controller regarding the vision system and the vehicle controller, indicating quantitative features (see Table 3 and Section 6.2). Moreover, algorithms to allow the robots in the fleet to collaborate, follow a plan, and avoid collisions between robots by using the master-slave configuration have been presented in Section 6.3. In general, the proposed system has been assessed as very efficient to easily integrate new sensors, implements, and innovative algorithms in a fleet of agricultural robots.

The industrial exploitation of the fully unmanned fleet concepts presented in this paper is not yet permitted by the legislation of most countries. Nevertheless, the use of autonomous vehicles on public roads is under consideration in Japan, Sweden, and several states in the USA, and autonomous cars will unquestionably be allowed everywhere in the near future. In any case, the authorization of autonomous vehicles for closed scenarios such as farms will definitely occur first, and researchers are preparing for this eventuality.

CONFLICT OF INTERESTS

The authors declare that there is no conflict of interests regarding the publication of this paper.

ACKNOWLEDGMENT

The research leading to these results has received funding from the European Union's Seventh Framework Programme [FP7/2007-2013] under Grant Agreement no. 245986.

REFERENCES

1. B. Åstrand and A.-J. Baerveldt, "An agricultural mobile robot with vision-based perception for mechanical weed control," Autonomous Robots, vol. 13, no. 1, pp. 21–35, 2002.

2. T. Bakker, K. van Asselt, J. Bontsema, J. Müller, and G. van Straten, "Autonomous navigation using a robot platform in a sugar beet field," Biosystems Engineering, vol. 109, no. 4, pp. 357–368, 2011.

3. M. Li, K. Imou, K. Wakabayashi, and S. Yokoyama, "Review of research on agricultural vehicle autonomous guidance," International Journal of Agricultural and Biological Engineering, vol. 2, no. 3, pp. 1–16, 2009.

4. S. M. Pedersen, S. Fountas, H. Have, and B. S. Blackmore, "Agricultural robots—system analysis and economic feasibility," Precision Agriculture,

vol. 7, no. 4, pp. 295–308, 2006.

5.   A. Stentz, C. Dima, C. Wellington, H. Herman, and D. Stager, "A system for semi-autonomous tractor operations," Autonomous Robots, vol. 13, no. 1, pp. 87–104, 2002.

6.   S. Gan-Mor, R. L. Clark, and B. L. Upchurch, "Implement lateral position accuracy under RTK-GPS tractor guidance," Computers and Electronics in Agriculture, vol. 59, no. 1-2, pp. 31–38, 2007.

7.   A. T. Nieuwenhuizen, J. W. Hofstee, and E. J. van Henten, "Performance evaluation of an automated detection and control system for volunteer potatoes in sugar beet fields," Biosystems Engineering, vol. 107, no. 1, pp. 46–53, 2010.

8.   M. Pérez-Ruiz, D. C. Slaughter, C. J. Gliever, and S. K. Upadhyaya, "Automatic GPS-based intra-row weed knife control system for transplanted row crops," Computers and Electronics in Agriculture, vol. 80, pp. 41–49, 2012.

9.   D. C. Slaughter, D. K. Giles, and D. Downey, "Autonomous robotic weed control systems: a review,"Computers and Electronics in Agriculture, vol. 61, no. 1, pp. 63–78, 2008.

10.   L. Tian, "Development of a sensor-based precision herbicide application system," Computers and Electronics in Agriculture, vol. 36, no. 2-3, pp. 133–149, 2002.

11.   N. D. Tillett, T. Hague, A. C. Grundy, and A. P. Dedousis, "Mechanical within-row weed control for transplanted crops using computer vision," Biosystems Engineering, vol. 99, no. 2, pp. 171–178, 2008. ·

12.   F. Rovira-Más, "Sensor architecture and task classification for agricultural vehicles and environments,"Sensors, vol. 10, no. 12, pp. 11226–11247, 2010.

13.   T. Bakker, K. van Asselt, J. Bontsema, J. Müller, and G. van Straten, "A path following algorithm for mobile robots," Autonomous Robots, vol. 29, no. 1, pp. 85–97, 2010.

14.   M. Nørremark, H. W. Griepentrog, J. Nielsen, and H. T. Søgaard, "The development and assessment of the accuracy of an autonomous GPS-based system for intra-row mechanical weed control in row crops," Biosystems Engineering, vol. 101, no. 4, pp. 396–410, 2008.

15.   Y. Nagasaka, H. Saito, K. Tamaki, M. Seki, K. Kobayashi, and K. Taniwaki, "An autonomous rice transplanter guided by global positioning system and inertial measurement unit," Journal of Field Robotics, vol. 26, no. 6-7, pp. 537–548, 2009.

16. A. Bautin, O. Simonin, and F. Charpillet, "Towards a communication free coordination for multi-robot exploration," in Proceedings of the 6th National Conference on Control Architectures of Robots, pp. 8–14, Grenoble, France, May 2011.

17. N. Bouraqadi, L. Fabresse, and A. Doniec, "On fleet size optimization for multi-robot frontier-based exploration," in Proceedings of the 7th National Conference on Control Architectures of Robots, Nancy, France, May 2012.

18. P. Cartade, R. Lenain, B. Thuilot, B. Benet, and M. Berducat, "Motion control of a heterogeneous fleet of mobile robots: formation control for achieving agriculture task," in Proceedings of the International Conference on Agricultural Engineering (CIGR-AgEng ‹12), Valencia, Spain, July 2012.

19. RHEA, "A robot fleet for highly effective agriculture and forestry management," 2013, http://www.rhea-project.eu.

20. B. S. Blackmore, W. Stout, M. Wang, and B. Runov, "Robotic agriculture—the future of agricultural mechanisation?" in Proceedings of the 5th European Conference on Precision Agriculture, pp. 621–628, Uppsala, Sweden, June 2005.

21. D. Peleg, "Distributed coordination algorithms for mobile robot swarms: new directions and challenges," in Proceedings of the 7th International Workshop on Distributed Computing (IWDC ‹05), A. Pal, A. Kshemkalyani, R. Kumar, and A. Gupta, Eds., Springer, 2005.

22. B. K.-S. Cheung, K. L. Choy, C.-L. Li, W. Shi, and J. Tang, "Dynamic routing model and solution methods for fleet management with mobile technologies," International Journal of Production Economics, vol. 113, no. 2, pp. 694–705, 2008.

23. D. D. Bochtis, C. G. Sørensen, and S. G. Vougioukas, "Path planning for in-field navigation-aiding of service units," Computers and Electronics in Agriculture, vol. 74, no. 1, pp. 80–90, 2010.

24. B. S. Blackmore, H. W. Griepentrog, H. Nielsen, M. Nørremark, and J. Resting-Jeppesen, "Development of a deterministic autonomous tractor," in Proceedings of the CIGR International Conference, Beijing, China, November 2004.

25. E. Wisserodt, J. Grimm, M. Kemper et al., "Gesteuerte Hacke zur Beikrautregulierung innerhalb der Reihe von. Pflanzenkulturen. [Controlled Hoe for Weeding within Crop Rows]," in Tagung Landtechnik 1999/VDI-MEG, pp. 155–160, VDI-Verlag, Düsseldorf, Germany, 1999.

26. J. F. Reid, Q. Zhang, N. Noguchi, and M. Dickson, "Agricultural automatic

guidance research in North America," Computers and Electronics in Agriculture, vol. 25, no. 1-2, pp. 155–167, 2000.

27. T. Hinterhofer and S. Tomic, "Wireless QoS-enabled multi-technology communication for the RHEA robotic fleet," in Proceedings of the 1st Workshop on Robotics and Associated High-technologies and Equipment for Agriculture (RHEA ‹11›), P. Gonzalez-de-Santos and G. Rabatel, Eds., pp. 173–186, Montpellier, France, September 2011.

28. J. Carballido, M. Perez-Ruiz, C. Gliever, and J. Agüera, "Design, development and lab evaluation of a weed control sprayer to be used in robotic systems," in Proceeding of the 1st International Conference of Robotics and associated High-technologies and Equipment for agriculture (RHEA ‹12›), pp. 23–29, Pisa, Italy, September 2012.

29. A. Peruzzi, C. Frasconi, L. Martelloni, M. Fontanelli, and M. Raffaelli, "Application of precision flaming to maize and garlic in the RHEA project," in Proceedings of the 1st International Conference on Robotics and associated High-Technologies and Equipment for Agriculture (RHEA ‹12›), pp. 55–60, Pisa, Italy, September 2012.

30. M. Vieri, R. Lisci, M. Rimediotti, and D. Sarri, "The innovative RHEA airblast sprayer for tree crop treatment," in Proceedings of the 1st International Conference on Robotics and Associated High-Technologies and Equipment for Agriculture (RHEA ‹12›), pp. 93–98, Pisa, Italy, September 2012.

31. MOOS, "Cross Platform Software for Robotics Research," 2013,http://www.robots.ox.ac.uk/~mobile/MOOS/wiki/pmwiki.php/Main/HomePage.

32. PLAYER, "Free Software tools for robot and sensor applications," 2013,http://playerstage.sourceforge.net.

33. CARMEN, "The Carnegie Mellon Robot Navigation Toolkit," 2013,http://carmen.sourceforge.net/intro.html.

34. ROS, "Robot Operating System," 2013, http://www.ros.org.

35. R. Stallman, "The BSD License Problem," Free Software Foundation, 2013,http://web.archive.org/web/20061112224151/http://www.gnu.org/philosophy/bsd.html.

36. S. Cousins, "Is ROS good for robotics?" IEEE Robotics and Automation Magazine, vol. 19, no. 1, pp. 13–14, 2012.

37. F. Tobe, "IEEE Spectrum Blog," 2013, http://spectrum.ieee.org/automaton/robotics/robotics-software/irobot-willow-garage-debate.

38. P. Courrier, J. Hurdus, C. Reinholtz, and A. Wicks, "Team Victor Tango's

Odin: Autonomous Driving Using NI LabVIEW in the DARPA Urban Challenge, NI Case Study," 2012,http://sine.ni.com/cs/app/doc/p/id/cs-11323.

39. J. M. Ramírez, P. Gómez-Gil, and F. L. Larios, "A robot-vision system for autonomous vehicle navigation with fuzzy-logic control using labview," in Proceedings of the Electronics, Robotics and Automotive Mechanics Conference (CERMA ‹07), pp. 295–300, Cuernavaca, México, September 2007. ·

40. P. Gonzalez-de-Santos, A. Ribeiro, and C. Fernandez-Quitanilla, "The RHEA Project: using a robot fleet for a highly effective crop protection," Proceedings of the International Conference of Agricultural Engineering (CIGR-Ageng ‹12), Valencia, Spain, July 2012.

41. D. D. Bochtis and C. G. Sørensen, "The vehicle routing problem in field logistics part I," Biosystems Engineering, vol. 104, no. 4, pp. 447–457, 2009.

42. D. D. Bochtis and C. G. Sørensen, "The vehicle routing problem in field logistics: part II," Biosystems Engineering, vol. 105, no. 2, pp. 180–188, 2010.

43. R. Eaton, J. Katupitiya, K. W. Siew, and K. S. Dang, "Precision guidance of agricultural tractors for autonomous farming," in Proceedings of the 2nd Annual IEEE Systems Conference Proceedings (SysCon ‹08), pp. 314–321, April 2008.

44. C. G. Sørensen and D. D. Bochtis, "Conceptual model of fleet management in agriculture," Biosystems Engineering, vol. 105, no. 1, pp. 41–50, 2010.

45. S. G. Vougioukas, "A distributed control framework for motion coordination of teams of autonomous agricultural vehicles," Biosystems Engineering, vol. 113, pp. 284–297, 2012.

46. D. A. Johnson, D. J. Naffin, J. S. Puhalla, J. Sanchez, and C. K. Wellington, "Development and implementation of a team of robotic tractors for autonomous peat moss harvesting," Journal of Field Robotics, vol. 26, no. 6-7, pp. 549–571, 2009.

47. S. J. Moorehead, C. K. Wellington, B. J. Gilmore, and C. Vallespi, "Automating orchards: a system of autonomous tractors for orchard maintenance," in Proceedings of the IEEE International Conference of Intelligent Robots and Systems, Workshop on Agricultural Robotics, 2012.

48. V. Arguenon, A. Bergues-Lagarde, P. Bro, C. Rosenberger, and W. Smari, "Multi-agent based prototyping of agriculture robots," in Proceedings of the International Symposium on Collaborative Technologies and Systems

(CTS ‹06), pp. 282–288, May 2006.·

49. Y. Hao, B. Laxton, E. R. Benson, and S. K. Agrawal, "Differential flatness-based formation following of a simulated autonomous small grain harvesting system," Transactions of the American Society of Agricultural Engineers, vol. 47, no. 3, pp. 933–944, 2004.

50. N. Noguchi, J. Will, J. Reid, and Q. Zhang, "Development of a master-slave robot system for farm operations," Computers and Electronics in Agriculture, vol. 44, no. 1, pp. 1–19, 2004.·

51. R. A. Brooks, "A robust layered control system for a mobile robot," IEEE Journal of Robotics and Automation, vol. 2, no. 1, pp. 14–23, 1986.

52. D. Payton, "An architecture for reflexive autonomous vehicle control," in Proceedings of the IEEE International Conference on Robotics and Automation, pp. 1838–1845, San Francisco, Calif, USA, February 1986.

53. D. Nakhaeinia, S. H. Tang, S. B. Mohd Noor, and O. Motlagh, "A review of control architectures for autonomous navigation of mobile robots," International Journal of Physical Sciences, vol. 6, no. 2, pp. 169–174, 2011.

54. L. Emmi, G. Pajares, and P. Gonzalez-de-Santos, "Integrating robot positioning controllers in the SEARFS simulation environment," in Proceedings of the 1st International Conference on Robotics and associated High-Technologies and Equipment for Agriculture (RHEA ‹12), pp. 151–156, Pisa, Italy, September 2012.

55. S. I. Cho and J. H. Lee, "Autonomous speedsprayer using differential global positioning system, genetic algorithm and fuzzy control," Journal of Agricultural Engineering Research, vol. 76, no. 2, pp. 111–119, 2000.

56. T. Hague, J. A. Marchant, and N. D. Tillett, "Ground based sensing systems for autonomous agricultural vehicles," Computers and Electronics in Agriculture, vol. 25, no. 1-2, pp. 11–28, 2000.

57. V. Subramanian, T. F. Burks, and A. A. Arroyo, "Development of machine vision and laser radar based autonomous vehicle guidance systems for citrus grove navigation," Computers and Electronics in Agriculture, vol. 53, no. 2, pp. 130–143, 2006.

58. J. Xue, L. Zhang, and T. E. Grift, "Variable field-of-view machine vision based row guidance of an agricultural robot," Computers and Electronics in Agriculture, vol. 84, pp. 85–91, 2012.

59. J. Blasco, N. Aleixos, J. M. Roger, G. Rabatel, and E. Moltó, "Robotic weed control using machine vision," Biosystems Engineering, vol. 83, no. 2, pp. 149–157, 2002.

60.   W. S. Lee, D. C. Slaughter, and D. K. Giles, "Robotic weed control system for tomatoes," Precision Agriculture, vol. 1, no. 1, pp. 95–113, 1999.

61.   V. Leemans and M.-F. Destain, "A computer-vision based precision seed drill guidance assistance,"Computers and Electronics in Agriculture, vol. 59, no. 1-2, pp. 1–12, 2007.

CITATION

CHAPTER 1

Lei Wang, Muguo Li, Junyan Qi, and Qun Zhang, "Design Approach Based on EtherCAT Protocol for a Networked Motion Control System," International Journal of Distributed Sensor Networks, vol. 2014, Article ID 750601, 15 pages, 2014. doi:10.1155/2014/750601

CHAPTER 2

D. Fisher and P. Gould, "Open-Source Hardware Is a Low-Cost Alternative for Scientific Instrumentation and Research," Modern Instrumentation, Vol. 1 No. 2, 2012, pp. 8-20. doi: 10.4236/mi.2012.12002.

CHAPTER 3

Vogel-Heuser, B. , Rösch, S. , Fischer, J. , Simon, T. , Ulewicz, S. and Folmer, J. (2016) Fault Handling in PLC-Based Industry 4.0 Automated Production Systems as a Basis for Restart and Self-Configuration and Its Evaluation. Journal of Software Engineering and Applications, 9, 1-43. doi: 10.4236/jsea.2016.91001.

CHAPTER 4

Giap, N. , Shin, J. and Kim, W. (2014) A Reference-Pulse Generator for Motion Control System. Intelligent Control and Automation, 5, 111-119. doi: 10.4236/ica.2014.53013.

CHAPTER 5

Thinh, N. , Choi, J. and Kim, W. (2014) Design and Implementation of FFPIV Scheme for Closed Loop Motion Controller. Intelligent Control and Automation, 5, 35-45. doi: 10.4236/ica.2014.52005.

CHAPTER 6

Timo Vepsäläinen and Seppo Kuikka, "Model-Driven Development of Automation and Control Applications: Modeling and Simulation of Control Sequences,"Advances in Software Engineering, vol. 2014, Article ID 470201, 14 pages, 2014. doi:10.1155/2014/470201

CHAPTER 7

Muhammad Mohsin Khan, Asad Ullah Awan, and Muwahida Liaquat, "Improving Vehicle Handling and Stability under Uncertainties using Probabilistic Approach," IFAC-PapersOnLine, vol. 48, no. 25, pp. 242–247, 2015.

CHAPTER 8

Erwin Normanyo, Francis Husinu, Ofosu Robert Agyare; Developing a Human Machine Interface (HMI) for Industrial Automated Systems using Siemens Simatic WinCC Flexible Advanced Software; ISSN 2079-8407S

CHAPTER 9

Ryan A. Beasley, "Medical Robots: Current Systems and Research Directions," Journal of Robotics, vol. 2012, Article ID 401613, 14 pages, 2012. doi:10.1155/2012/401613

CHAPTER 10

Luis Emmi, Mariano Gonzalez-de-Soto, Gonzalo Pajares, and Pablo Gonzalez-de-Santos, "New Trends in Robotics for Agriculture: Integration and Assessment of a Real Fleet of Robots," The Scientific World Journal, vol. 2014, Article ID 404059, 21 pages, 2014. doi:10.1155/2014/404059

INDEX

3